Handbook of
Turbomachinery and
Mechanical Engineering

Handbook of Turbomachinery and Mechanical Engineering

Contributors

Ariavie Go, Oyekale Jo et al.

AURIS
Reference

www.aurisreference.com

Handbook of Turbomachinery and Mechanical Engineering

Contributors: Ariavie Go, Oyekale Jo et al.

Published by Auris Reference Limited

www.aurisreference.com

United Kingdom

Copyright 2016
Printed in 2017 for Sale in the Indian Subcontinent

Handbook of Turbomachinery and Mechanical EngineeringBest Practice

ISBN: 978-1-78154-947-6

British Library Cataloguing in Publication Data
A CIP record for this book is available from the British Library

Printed in the United Kingdom

Exclusively distributed by CBS Publishers & Distributors Pvt. Ltd.

Sales & Distribution Rights only for India, Pakistan, Bangladesh, Sri Lanka, Nepal and Bhutan.This book is not to be sold outside these territories.

Contents

List of Abbreviations

AIP	aerodynamic interface plane
AFTRF	Axial Flow Turbine Research Facility
CFD	Computational Fluid Dynamics
CFD	control and processing module"CPM
CMM	coordinate measuring machine
EO	engine order
FD	Finite Differences
FHP	Five-Hole Probes
FM	frequency modulated
GA	Genetic Algorithms
IGV	inlet guide vane
ITSM	Institute of Thermal Turbomachinery
IBRs	integrally bladed rotors
LOM	Largest of maximum
MCC	Measurement Computing Corporation
NL	nominal loading
NSMS	nonintrusive stress measurement systems,
NGV	nozzle guide vanes
OPR	once-per-revolution
RSS	root-sum-squared
RSI	Rotorstator interaction
SQP	Sequential Quadratic Programming
SA	Simulated Annealing
SOM	Smallest of maximum
TTL	transistor-transistor logic
FIS	Fuzzy Inference System

List of Contributors

Ariavie Go
Department of Mechanical Engineering

Oyekale Jo
University of Benin, Benin City Federal University of Petroleum Resources, Effurun

Emagbetere E
University of Benin, Benin City Federal University of Petroleum Resources, Effurun

Chih-Neng Hsu
Department of Refrigeration, Air Conditioning and Energy Engineering, National Chin-Yi University of Technology, Taichung City 41170, Taiwan

R A Van den Braembussche
Von Karman Institute for Fluid Dynamics, Waterloose Steenweg, 72, B-1640 SintGenesius-Rode, Belgium

X X Huang
Center for Industrial Diagnostics and Fluid Dynamics (CDIF), Universitat Politècnica de Catalunya, ETSEIB Building, Av. Diagonal 647 –08028 Barcelona (Spain)

E Egusquiza
Center for Industrial Diagnostics and Fluid Dynamics (CDIF), Universitat Politècnica de Catalunya, ETSEIB Building, Av. Diagonal 647 –08028 Barcelona (Spain)

C Valero
Center for Industrial Diagnostics and Fluid Dynamics (CDIF), Universitat Politècnica de Catalunya, ETSEIB Building, Av. Diagonal 647 –08028 Barcelona (Spain)

A Presas
Center for Industrial Diagnostics and Fluid Dynamics (CDIF), Universitat Politècnica de Catalunya, ETSEIB Building, Av. Diagonal 647 –08028 Barcelona (Spain)

Cheng Xu
Department of Mechanical Engineering, University of Wisconsin-Milwaukee,
Milwaukee, WI 53212, USA

Ryoichi S. Amano
Department of Mechanical Engineering, University of Wisconsin-Milwaukee,
Milwaukee, WI 53212, USA

Roberto Capata
Department of Mechanical and Aerospace Engineering, University of Roma
"Sapienza", Roma 00184, Italy

Enrico Sciubba
Department of Mechanical and Aerospace Engineering, University of Roma
"Sapienza", Roma 00184, Italy

Masami Suzuki
Department of Mechanical Systems Engineering, University of the Ryukyus,
Okinawa, Japan

F. R. Menter
ANSYS GmbH Germany

R. B. Langtry
The Boeing Company,USA

Jason Town
Turbomachinery Aero-Heat Transfer Laboratory, Department of Aerospace
Engineering, The Pennsylvania State University, University Park, PA 16802,
USA

Cengiz Camci
Turbomachinery Aero-Heat Transfer Laboratory, Department of Aerospace
Engineering, The Pennsylvania State University, University Park, PA 16802,
USA

William L. Murray III
Purdue University, 500 Allison Road, West Lafayette, IN 47907, USA

Nicole L. Key
Purdue University, 500 Allison Road, West Lafayette, IN 47907, USA

Reid A. Berdanier
School of Mechanical Engineering, Purdue University, 500 Allison Road,
West Lafayette, IN 47907, USA

Nicole L. Key
School of Mechanical Engineering, Purdue University, 500 Allison Road,
West Lafayette, IN 47907, USA

Ravirai Jangir
Thermal Turbomachines Laboratory, Department of Mechanical Engineering,
IIT Madras, Chennai 600 036, India

Nekkanti Sitaram
Thermal Turbomachines Laboratory, Department of Mechanical Engineering,
IIT Madras, Chennai 600 036, India

Ct Gajanan
Thermal Turbomachines Laboratory, Department of Mechanical Engineering,
IIT Madras, Chennai 600 036, India

Steffen K̈ammerer
Institute of Thermal Turbomachinery and Machinery Laboratory, University
of Stuttgart, Stuttgart, Germany

J̈urgen F. Mayer
Institute of Thermal Turbomachinery and Machinery Laboratory, University
of Stuttgart, Stuttgart, Germany

Heinz Stetter
Institute of Thermal Turbomachinery and Machinery Laboratory, University
of Stuttgart, Stuttgart, Germany

Meinhard Paffrath
Siemens AG, Corporate Technology, Munich, Germany

Utz Wever
Siemens AG, Corporate Technology, Munich, Germany

Alexander R. Jung
Siemens AG Power Generation Group, Mulheim an der Ruhr, Germany

Preface

The text *Handbook of Turbomachinery and Mechanical Engineering* presents new material on advances in fluid mechanics of turbomachinery, high-speed, rotating, and transient experiments, cooling challenges for constantly increasing gas temperatures, advanced experimental heat transfer and cooling effectiveness techniques, and propagation of wake and pressure disturbances. Performance modelling of steam turbine performance using fuzzy logic membership functions has been focused in first chapter. A study on fluid self-excited flutter and forced response of turbomachinery rotor blade has been presented in second chapter. Third chapter discusses some design systems for turbomachinery applications that have been developed over the years in order to assist the designer in finding the optimal geometry by making better use of the available information. In fourth chapter, the dynamic response analysis for circular disks with different dimensions and diskblades-disk structures have been carried out to better understand the fundamental dynamic behavior for the complex turbomachinery. Fifth chapter provides some empirical information for designing industrial centrifugal compressors with a focus on the impeller. The purpose of sixth chapter is to present and discuss a preliminary and simple method to extend the currently available design maps into the small scale range ($Re < 10^5$) by introducing in the Balje charts an efficiency correction that depends on the specific speed n_s. Numerical analysis of horizontal-axis wind turbine characteristics in yawed conditions has been presented in seventh chapter. In eight chapter, a novel approach to simulating laminar to turbulent transition is described that can be implemented into a general RANS environment. Ninth chapter focuses on time efficient adaptive gridding approach and improved calibrations in five-hole probe measurements. Tenth chapter explains the particular data processing methods used to identify rotor vibration. Experimental investigation of factors influencing operating rotor tip clearance in multistage compressors has been performed in eleventh chapter. A miniature four-hole probe with minimum spatial error has been designed and fabricated in last chapter.

Chapter 1

PERFORMANCE MODELLING OF STEAM TURBINE PERFORMANCE USING FUZZY LOGIC MEMBERSHIP FUNCTIONS

Ariavie, Go[1]; Oyekale Jo[2] Emagbetere E[2]

[1]Department of Mechanical Engineering

[2]University of Benin, Benin City Federal University of Petroleum Resources, Effurun

ABSTRACT

A Fuzzy Inference System for predicting the performance of steam turbine based on Rankine cycle is developed using a 144-rule based in analyzing the generated data for different inlet and outlet conditions. The result of efficiency for different types of membership functions and defuzzification method was obtained. Centroid method of defuzzification gave good results irrespective of the type of membership function with error less than 5%. However, other defuzzification methods gave good result for some types of membership functions. Result of different input data tested do not vary significantly (P<

INTRODUCTION

Performance prediction tends to estimate useful parameters such as efficiency based on certain input parameters (Benner, Sjolander, and Moustapha, 2006a, Benner, Sjolander, and Moustapha, 2006b, Benner, Sjolander, and Moustapha, 2004, Dunham and Came, 1970, Živković, 2000). Steam turbine is mostly used for generating mechanical energy due to its advanced features and outstanding qualities (Mathis, 2003). Currently, researchers are working assiduously towards improving efficiency of steam turbine by developing better metallurgical materials that can withstand steam at temperature as high as 6000C and beyond critical pressure (Brooks,2012). However, not much improvement on its efficiency has been observed in recent times, due to the long technological history of efficiency of turbine cycles being about 52%, which is about 30% higher than that of a simple gas cycle, and about 40% as reported by General Electric (Organoski, 1990). One challenge faced by end-users is turbine operating at off-design conditions almost throughout its life, therefore, manufacturers need to accurately predict performance at varying operating conditions and incorporate a control system performance monitoring. Performance prediction has graduated from qualitative guess checks to several empirical, analytical and numerical methods. Empirical methods suitable for gas turbines (Ainley and Mathieson, 1951, Dunham and Came, 1970, and Craig and Cox, 1971) can be modified with suitable correlations for non-condensing steam turbine modeling (Denton, 1993). However, Dixon (2005) argued that the empirical methods lack first principle in their derivations, and he developed a model that cannot be implemented because it requires quantifying all losses which is practically impossible. Existing analytical methods in standard literature (Horlock and Denton, 2003) come with rigorous mathematical complexities and require requisite knowledge of turbo-machinery physics for their application. There are well developed numerical methods such as Computational Fluid Dynamics (CFD), but they are very expensive and time consuming. Besides, their results are not accurate as reported by Denton (Emami, 1997). Current market trend of steam turbines has necessitated the development of easy to use performance prediction methods that give good result. Methods that do not require mathematical formulation such as Fuzzy logic, genetic algorithm and artificial neural network are therefore good alternatives.

Fuzzy Logic is one of the best methods that do not require mathematical formulation. It replaces rigorous, mathematical formulations with ambiguous qualitative and random conceptions (Ariavie, Ovuworie and Ariavie, 2011). It is easy to use and has been applied to various systems ranging from linear to high level non-linear systems. Fuzzy logic tends to map a set of inputs to a set of outputs, obeying a list of rules (combination of statements with inference). The rules are formed from the variables (members of input and output sets) and the adjectives that describe those variables (membership functions). Before one can build a system that interprets rules, it is imperative to define all the terms to be used and the adjectives that describe them. To say that the pressure is high, one needs to define the range that the steam pressure can be expected to vary as well as what is implied by the word high (Harris, 2006). Due to its advantages and wide applicability, the method has gained much development and is discussed in detail as a subject in standard literature (Cox, 1994, Perfilieva and Mockr, 1999, Sivanandam et al., 2001, Yager and Filer,1994, Micheal and Shapiro, 2006). This work investigated the application of Fuzzy Logic to performance prediction of steam turbine. A Fuzzy Inference System (FIS) was developed. The FIS can be used to determine the thermodynamic efficiency of a steam turbine from the steam inlet pressure, inlet temperature and exit pressure as input variables. The FIS was trained with analyzed data generated from developed MATLAB simulation code that estimates efficiency, back work ratio and specific steam consumption based on the Rankine Cycle. The effect of different types of membership functions and defuzzification methods available in the MTALB FIS development environment was investigated and compared to the analytical result.

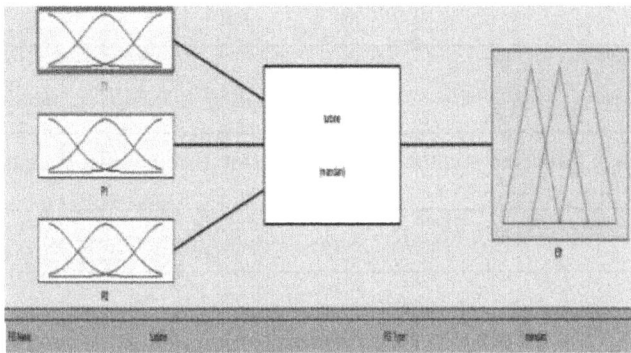

Figure .1: Mandini Fuzzy Inference System for Performance Prediction of Steam Turbine

METHODS AND MATERIALS

Data was generated using Matlab simulation code for steam turbine modeling; however some suitable assumptions were made. The data was analyzed, filtered and then used to train the fuzzy inference systems developed for steam turbine performance predictions. MATLAB Simulation program based on the Rankine combine cycle for steam turbine modeling was developed. Basic assumptions for applying the Rankine cycle to steam turbine modeling given in standard literature (Celgel, and Boles,2006, Potter and Somerton, 2004) were made. Optimum reheat temperature of one-fourth inlet conditions (Celgel, and Boles, 2006) was adopted, and varying conditions were adopted for the regenerative cycle. Turbine and pump isentropic efficiency were taken as 85%. The code was used to simulate result for varying turbine conditions. The data was analyzed in order to ascertain the rules for the fuzzy inference system. Turbine inlet and outlet conditions (pressure and temperature), and their effects on the turbine efficiency were investigated. Development of Fuzzy Inference System: The Mandini type of Fuzzy Inference System was adopted for this work. The turbine inlet conditions (Boiler pressure (P1) and temperature (T1) respectively) and the turbine exit pressure (Condenser pressure (P2)) were taken as the three inputs variables while the efficiency of the turbine was taken as the output of the Fuzzy Inference System (FIS) for performance prediction of steam turbine. This is shown in figure

The membership functions for each variable was designed with the following range of values as seen best for developing rules for the turbine based on the data analyzed.

Table 1: Range Values of membership function for Turbine inlet temperature (T1)

Values in degree Celsius	100 – 320	250 – 400	320 – 450	400 – 500	450 – 600
Designation	Very Low (VL)	Low (L)	Medium (M)	High (H)	Very High (VH)

Table 2: Range Values of membership function for Turbine Inlet Pressure (P1)

Values in MPa	1 – 4	2 – 8	5 – 12	7 – 18	12 – 20	16 – 24	
Designation	Very Low (VL)	Low (L)	Medium (M)	High (H)	Very High (VH)	Very High (VVH)	

Table 3: Turbine Exit Pressure (P2)

Values in MPa	0.0015 – 0.015	0.01 – 0.05	0.04 – 0.1	0.08 – 0.5	0.3 – 1.2
Designation	Very Low (VL)	Low (L)	Medium (M)	High (H)	Very High (VH)

Table .4: Range Values of membership function for Efficiency

Values	-0.2 – 0.2	0 – 0.4	0.2 – 0.6	0.4 – 0.8	0.6 – 1.0
Designation	Very Low (VL)	Low (L)	Medium (M)	High (H)	Very High (VH)

Four (4) different membership functions which includes the triangular (trimf), Gaussian (gausmf), Polynomial (Pmf) and the modified Gaussian (Gauss2mf) were tested with each generating its unique curve based on the function it describes. Their default parameters for each of the curves were adopted. The output (Efficiency) for each selected membership function was recorded. A sample dialog box for editing the type membership function and its parameters is shown in figure 2.2.

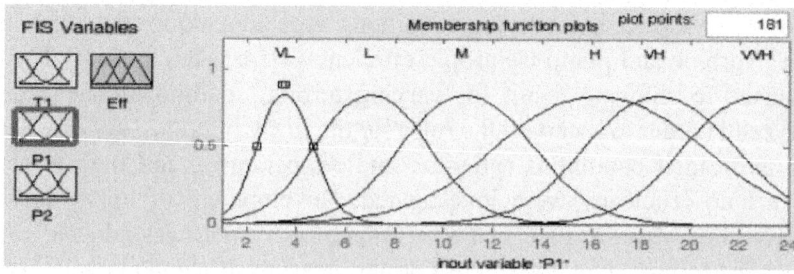

Figure 2: The Polynomial membership function (Pmf) for the Inlet temperature (T1)

Five (5) different defuzzification methods (the Centroid, Bisector of Area (BOA), Middle of maximum (MOM), Largest of maximum (LOM) and Smallest of maximum (SOM)) were also tested and corresponding values of efficiencies recorded. They include One hundred and forty four (144) rules were selected based on analysis of the generated data and a few is shown in figures 2.3 and 2.4, while Figures 2.5, 2.6, 2.7 and 2.8 shows the surface plots representing the rules. The minimum values and maximum values were used for the 'and' and 'or' combination respectively. The FIS developed was tested with the data below.

Table 2.5: Test data for fuzzy inference system

T1	P1	P2	Efficiency
360	4	0.05	0.371
600	24	0.02	0.502

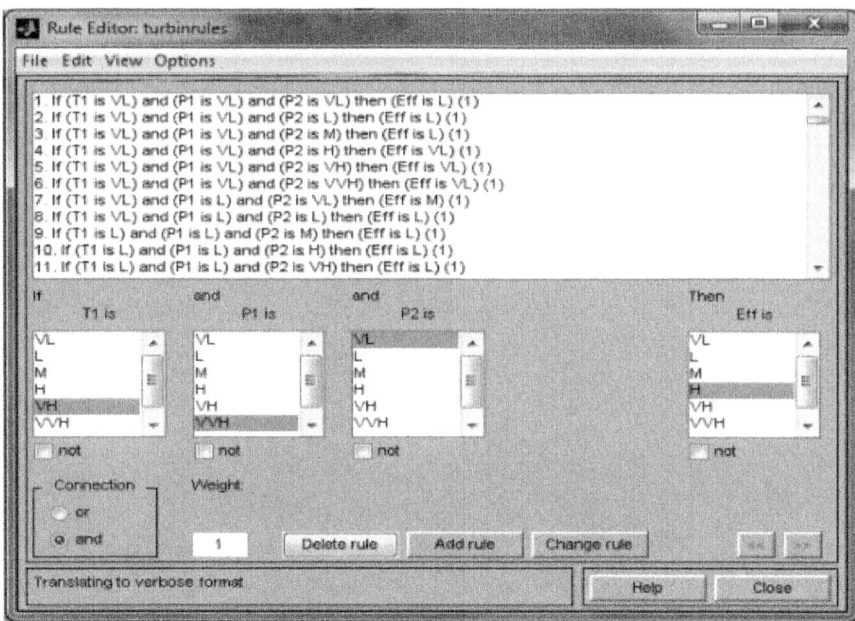

Figure. 3:

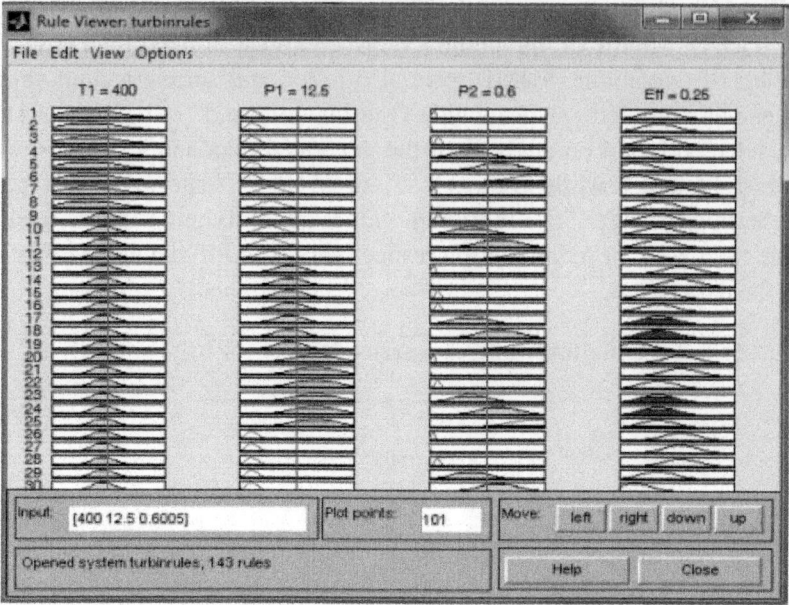

Figure.4:

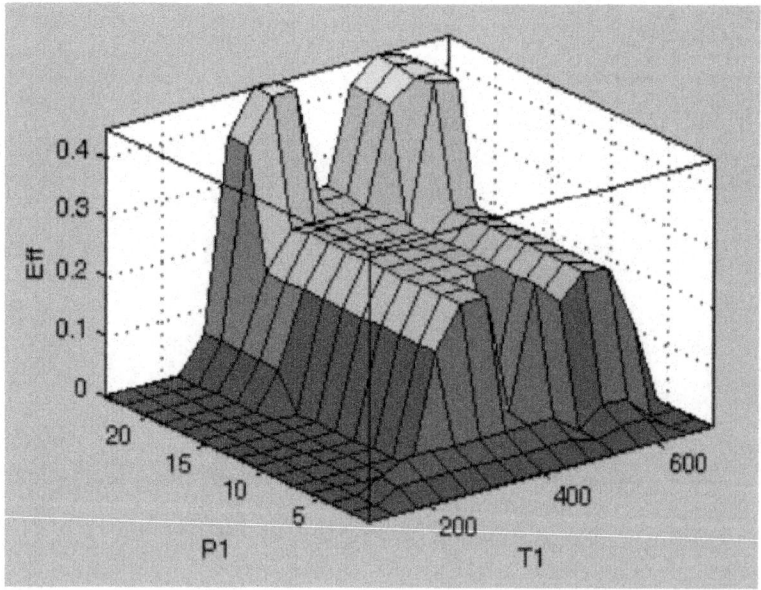

Figure.5:

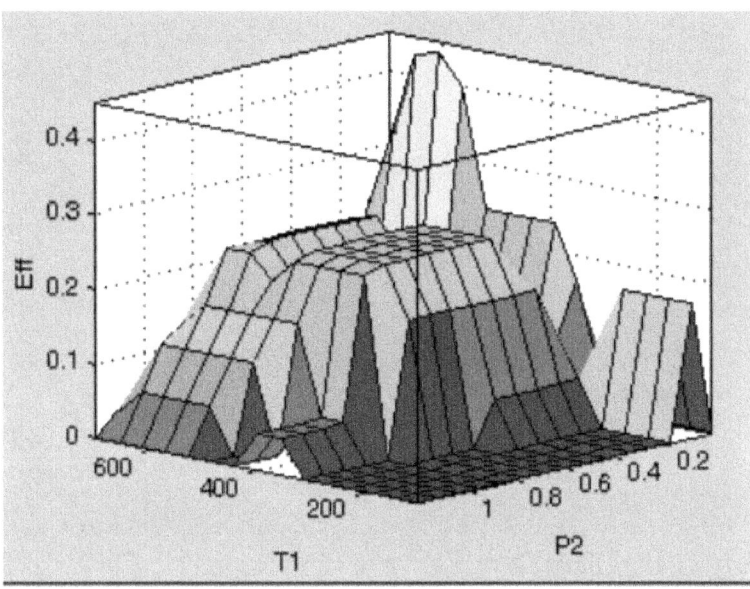

Figure.6:

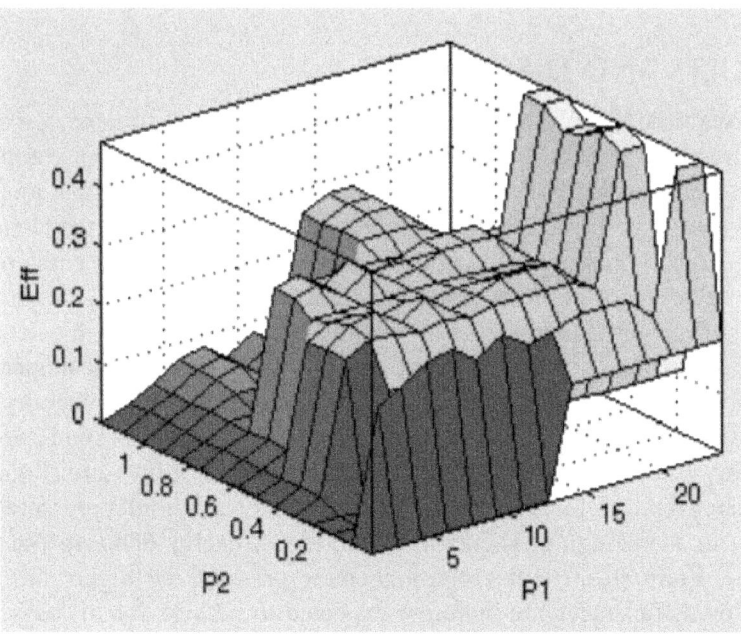

Figure.7:

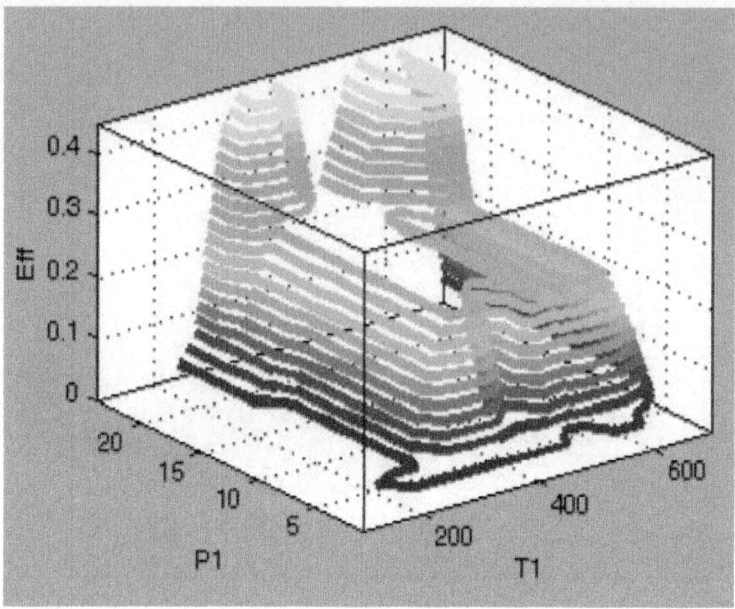

Figure.8:

RESULTS AND DISCUSSION

Data Analysis: Plot of Turbine exit pressure (P2) versus Efficiency for different inlet pressures (Figure 3.1) shows that efficiency tends to decrease nonlinearly as condenser pressure increases. This is explained by the fact that much more of the energy possessed by the body would have been converted to work at a more reduced pressure. This is further explained by specific enthalpy being lower values at lower pressure values and vice versa. Lower exit pressures give higher efficiency, and lower steam quality. There is a limit to which condenser pressure may be reduced since it affects the steam quality. For modern steam turbines design, steam at the exit of the turbine must be at least 90% dry. That is the dryness fraction which must be between 0.9 and 1 (Celgel and Boles, 2006) Contrary to decrease in efficiency with increasing exit pressure (Figure 3.1), efficiency tends to increase non-linearly with increasing inlet pressure (Figure 3.2). This increase in efficiency is due to higher energy of steam entering the turbine. From steam tables, the higher the pressure the higher the specific enthalpy. Efficiencies are therefore expected to be high for higher values of inlet pressure provided other conditions are kept constant as shown in Figure 3.1. There is a limit to which the pressure can be increased due to metallurgical strength for piping materials. However there are newly developed materials with strength high enough to withstand steam at critical pressure (Potter and Somerton, 2004). The efficiency of turbine increases linearly with inlet temperature of steam as shown in Fig. 3.3. This is however expected because heat is directly proportional to temperature. So steam at higher temperature possesses more heat energy that will be converted to work leading to higher efficiency values when other conditions are kept constant (Fig 3.2). So heating steam to high temperatures can increase the efficiency of the turbine, however this is however limited by the thermal properties of the material. There are modern developed materials that can withstand steam temperatures as high as 600oC (Potter and Somerton, 2004).

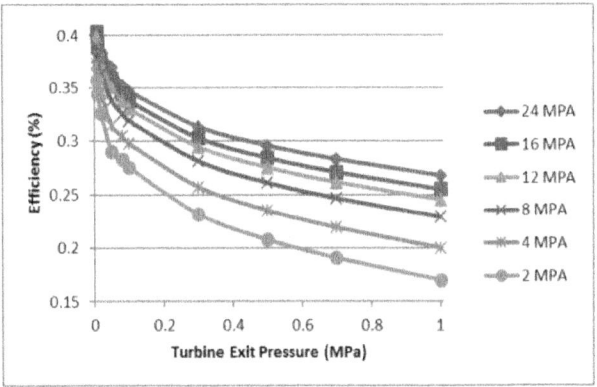

Figure.3: Plot of Steam exit pressure (MPa) versus efficiency for different inlet Pressure at 6000C

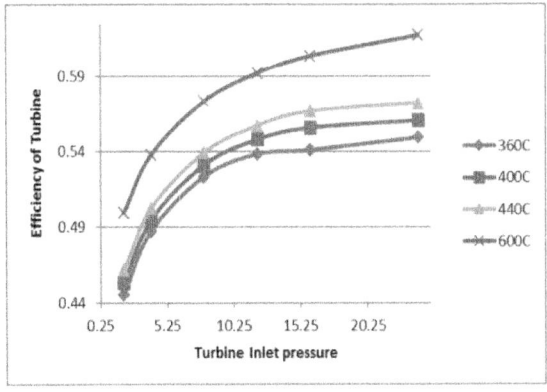

Figure.4: Effect of Boiler Pressure on the efficiency of Turbine for different Temperatures

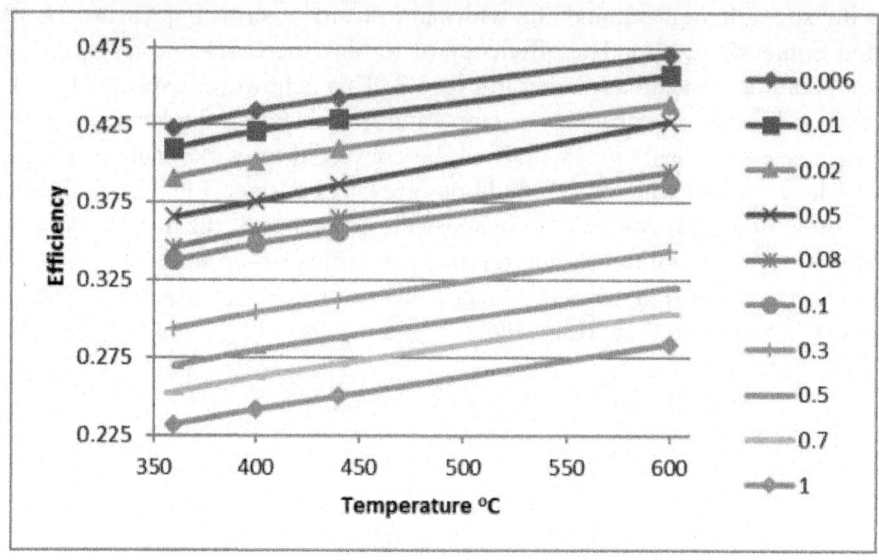

Figure. 5: Temperature versus Efficiency for Different Exit Pressure at 16MPa inlet pressure

Fuzzy Inference System: Figures 3.4 and 3.5 are bar charts showing the values of efficiency for different membership function and method of defuzzification. Their respective percentage error computed is shown in table 3.1. The Centroid method of defuzzification gives good result, that when compared to analytical values with efficiency less than 5%. The bisector method gives very good result for all membership function except the polynomial membership function in which the percentage error is about 7%. The other three methods of defuzzification: LOM, NOM and SOM respectively gave poor results irrespective of the type of membership function used, with their respective percentage error more than 5% for all cases. LOM method of defuzzification gave the worst result with percentage error of about 34% for triangular membership function. The triangular membership function with centroid method of defuzzification gave the best result with percentage error being about 0.5%, followed by the modified Gaussian method with percentage error of about 0.8%.

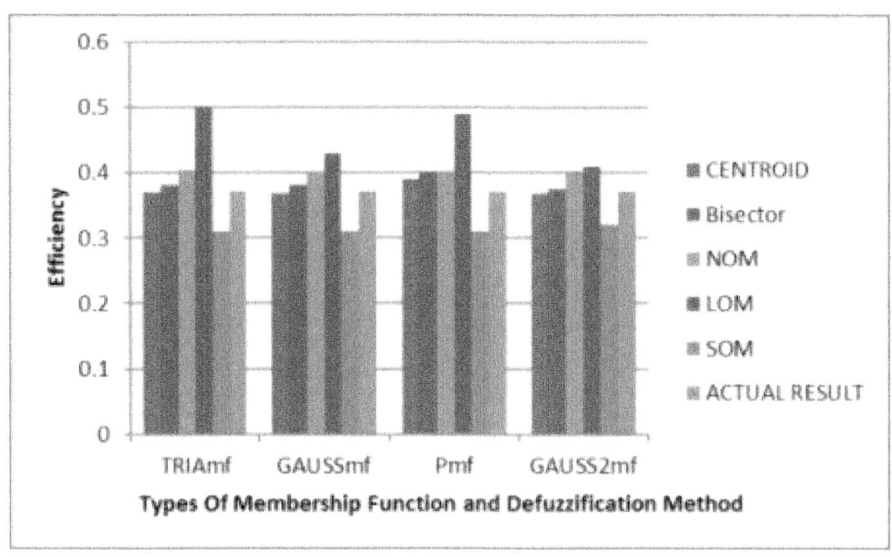

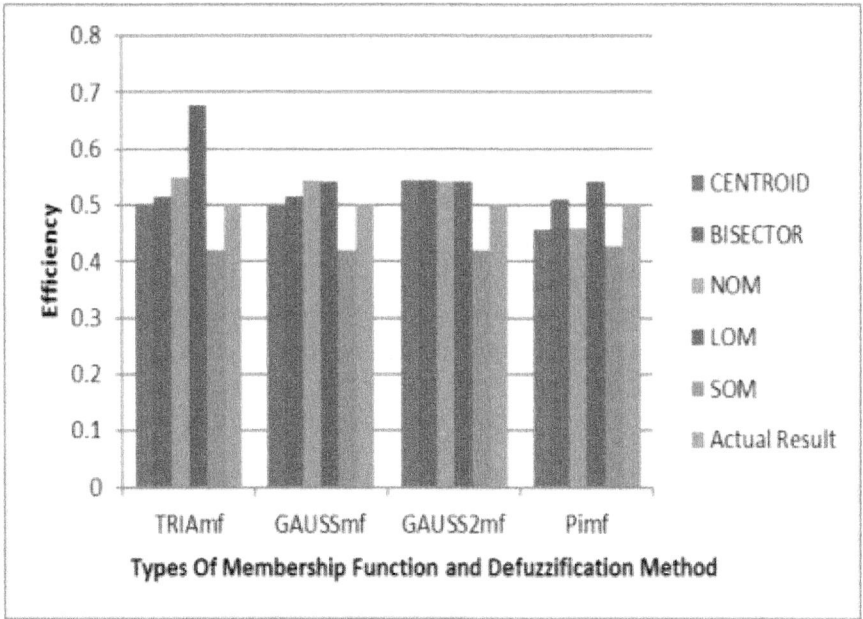

Figure 6: Chart of efficiency for different Membership functions and Defuzzification Methods compared to actual Efficiency for second case study.

Table 3.1: Efficiency computed for the different methods

	TRIAmf	GAUSSmf	Pmf	GAUSS2mf
CENTROID	0.539084	1.078167	5.012129	0.8086253
Bisector	2.425876	2.425876	7.816712	1.3477089
NOM	9.16442	7.816712	7.816712	7.5471698
LOM	34.77089	15.90296	32.07547	10.512129
SOM	16.44205	16.44205	16.44205	13.746631

Table 3.2: Statistical analysis of results

ANOVA							
Source	of	SS	df	MS	F	P-	F
Variation						value	critical
Rows		0.110784	24	0.0046	40.694	3.83E-	1.9837
				16	81	14	6
Columns		0.219085	1	0.2190	1931.4	1.89E-	4.2596
				85	55	24	77
Error		0.002722	24	0.0001			
				13			
Total		0.332591	49				

CONCLUSSION

In conclusion, Fuzzy logic can be effectively used for predicting performance of steam turbine. The efficiency of the Fuzzy Inference System depends on the method of defuzzification, and type of membership function and its parameter used in developing the system. However, the parameters of membership function and range can be adjusted to improve efficiency of the system.

REFERENCES

1.   Ainley, D.G., Mathieson, G.C.R. (1951) A method of performance estimation for axial-flow turbines, Aeronautical Research Council Reports & Memoranda, (2974), December 1951.

2.   Ariavie G.O, Ovuworie G.C and Ariavie S.S, (2011). Fuzzy Failure Probability of Transmission Pipelines in the Niger Delta Region of Nigeria:

3.   The Case of Third Party Activities. Journal of the Nigerian Association of Mathematical Physics. Vol. 18. (May 2011), pp 445 – 450.

4. Benner,M. W., Sjolander, S. A., Moustapha S. H. (2006a). An Empirical Prediction Method for Secondary Losses in Turbines |part i:

5. A New Loss Breakdown Scheme and Penetration Depth Correlation, Journal of Turbomachinery, 128(2):273-280.

6. Benner, M. W. Sjolander, S. A., Moustapha, S. H. (2006b). An Empirical Prediction Method for Secondary Losses in Turbines |part ii:

7. A New Secondary Loss Correlation, Journal of Turbomachinery, 128(2):281-291, 2006.

8. Benner, M. W.,Sjolander, S. A. and Moustapha, S. H.(2004). The Influence of leading-edge Geometry on Secondary Losses in a Turbine cascade at the design incidence, Journal of Turbomachinery, 126:277{287, 2004

9. Brooks, F.J. (2012) GE Gas Turbine Performance Characteristics. GER-3567H.pdf, Systems Schenetardy, NY.

10. Celgel, Y.A and Boles, M.A (2006). Thermodynamics: An Engineering Approach, McGraw-Hill, 2006.

11. Cox E (1994), The fuzzy Systems Handbook: A Practitioner's Guide to Building, Using and Maintaining Fuzzy Systems, Boston APProfessional, 1994.

12. Craig, H.R.M., Cox, H.J.A., (1971). Performance Estimation of Axial Flow Turbines. Proc. Inst.Mech. Engrs, 185:407-424, 1970-1971.

13. Denton, J.D (1993). Loss Mechanisms in Turbomachines. Journal of Turbomachinery, 115,1993.

14. Dixon, S.L. (2005) Fluid Mechanics and Thermodynamics of Turbomachinery, 5th edition, Boston: Elsevier-Butterworth-Heinemann, 2005.

15. Dunham, J., Came P.M, (1970). Improvements to the Ainley-Mathieson Method of Turbine

16. Performance Prediction, Journal of Engineering for Power, 92(3):252 - 256, 1970.

17. Emami, M. R. (1997). Systematic Methodology of Fuzzy-logic Modeling and Control and Application to Robotics. A thesis submitted in conformity with the requirements for the degree of Doctor of Philosophy, Graduate Department of Mechanical and Industrial Engineering University of Toronto, anada. Copyright by Mohamed Reza Emami 1997.

18. http://tspace.library.utoronto.ca/bitstream/1807/11   72/1/NQ28276.pdf (accessed 05/02/2015) Harris, J (2006). Fuzzy Logic Applications in Engineering Science, Int Series on Microprocessor-based and Intelligent System Engineering, Vol. 29, 2006.

19. Horlock, J.H., Denton, J.D (2003). A Review of Some Early Design Practice using Computational Fluid Dynamics and a Current Perspective. Journal of Turbomachinery, 127(1):5(13), 2003.

20. Mathis, D. M (2003). Handbook of Turbomachinery, chapter Fundamentals of Turbine Design, 2nd edition, New York: M. Dekker, 2003.

21. Micheal J. M, Shapiro H. N., (2006). Fundamental of Engineering Thermodynamics, England: John Wiley and sons Inc, ISBN-10 0-47003037-2, 2006

22. Organoski G. (1990) "Development of first 40% Thermal Efficiency Gas Turbine" presented at the 1990 ASME gas turbine congress (Brussels) GE marine and industrial engines service division, Ohio, June 1990.

23. Perfilieva, N. V., Mockr J. K. (1999), Mathematical Principles of Fuzzy Logic, Academic press, 1999. Potter, M.U and Somerton, C.W. (2004).

24. Thermodynamics for Engineers, Shaum Outline Series, ISBN-0-07-050707-4.e

25. Sivanandam S. N., Sumathi, S., Deepa, S. N (2001). Introduction to Fuzzy Logic using MATLAB, New York: Springer, Berlin Heidelberg, Springer, 2001.

26. Yager R. R. and Filer D. P., (1994)Essentials of Fuzzy Modeling and Control, New York: John Wiley, 1994.

27. Živković, D. (2000) "Nonlinear mathematical model of the condensing steam turbine", FACTA UNIVERSITATIS Series: Mechanical Engineering, Vol.1, No 7, pp. 871 – 878, , 2000.

28. UDC 621.134.5:519.87 Faculty of Mechanical Engineering, Beogradska 14, 18000.

Chapter 2

A STUDY ON FLUID SELF-EXCITED FLUTTER AND FORCED RESPONSE OF TURBOMACHINERY ROTOR BLADE

Chih-Neng Hsu

Department of Refrigeration, Air Conditioning and Energy Engineering, National Chin-Yi University of Technology, Taichung City 41170, Taiwan

ABSTRACT

Complex mode and single mode approach analyses are individually developed to predict blade flutter and forced response. These analyses provide a system approach for predicting potential aeroelastic problems of blades. The flow field properties of a blade are analyzed as aero input and combined with a finite element model to calculate the unsteady aero damping of the blade surface. Forcing function generators, including inlet and distortions, are provided to calculate the forced response of turbomachinery blading. The structural dynamic characteristics are obtained based on the blade mode shape obtained by using the finite element model. These approaches can provide turbine engine manufacturers, cogenerators, gas turbine generators, microturbine generators, and engine manufacturers with an analysis system to remedy existing flutter and forced response methods. The findings of this study can be widely applied to fans, compressors, energy turbine power plants, electricity, and cost saving analyses.

INTRODUCTION

The turbomachinery blade design has been extensively adopted in turbine engines, turbogenerators, microturbine generators, and cogenerators of fans, compressors, and turbine blades. However, excessive vibration due to flutters or forced responses often causes turbomachinery blade failure. Thus, engine manufacturers aim to prevent turbomachinery blade failures to achieve decreased development time and cost, lower maintenance cost, and fewer operational restrictions. One method of preventing blade failures is to increase blade structural damping by using either tip- or midspan shrouded blade designs.

Endurance is one of the most important considerations in turbomachinery blade design. Avoiding responsive blade resonance and preventing instability in turbomachinery are essential to the successful development and operation of gas turbine engines. Vibratory conditions produce stresses, which exceed allowable fatigue strength, reduce engine life, and in some cases even result in failure. Prior assessment of these responses followed by corresponding corrective actions ensures cost-effective designs and development effort.

Forced response is caused by vibration at levels that exceed material endurance limits, thereby causing high cycle fatigue failure. Blades vibrate in normal modes. Hence, a blade may have as many critical or maximum stress points as it has natural modes. The blade designer must determine the normal blade modes and calculate which mode has the greatest potential for resonance excitation. The source of stimuli is normally distorted in the flow to the rotor, which is caused by wakes shed by upstream struts or vanes and by separation of the upstream flow from the inlet. Separation of the upstream flow is normally precipitated by aircraft maneuver, gusts, cross wind, and, on occasion, ingestion of munitions exhaust gases.

REVIEW OF RELATED LITERATURE

Chiang and Kielb [1] presented a useful design tool to predict potential forced response, over and above the standard Campbell diagram approach. A fan inlet distortion is analyzed with measured distortion, and the predicted response agreed with the measured response. Chiang and Turner [2] developed an analysis system to predict the forced response of the compressor rotor blade caused by downstream stator vanes and struts. The description of the potential disturbance flow defect is obtained from a CFD model. The finite element method is used to provide the mode shapes and frequencies for the blade motion. Once structural damping is determined, the blade forced response is predicted by the system.

Murthy and Stefko [3] used the forced response prediction system, a software system, which integrates structural dynamic and steady and unsteady aerodynamic analyses to efficiently predict the forced dynamic stresses of turbomachinery blades to aerodynamic and mechanical excitations. The program also performs flutter analysis. Kielb and Chiang [4] described and assessed the current state of technology, providing examples of current research directions and defining research needs for flow defects, unsteady blade loads, and blade response in forced response analysis of the turbomachinery blade. Izsak and Chiang [5] presented prediction of wake strength as a key element in turbine and compressor forced response analysis. An empirical wake model and a 3D CFD flow solver are used and compared with wake data to assess the accuracy of the method. The empirical wake model predictions are compared with wake data obtained from a low-speed turbine, a compressor research facility, and a high-speed turbine facility. Izsak's paper provides a guide for applying empirical and CFD methods to model turbine and compressor wakes for blade forced response.

Manwaring and Wisler [6] developed a comprehensive series of experiments and analyses performed on compressor and turbine blading to evaluate the ability of current engineering/analysis models to predict unsteady aerodynamic loading of modern gas turbine blading. The predictions are experimentally compared, and their abilities are assessed to help guide designers in using these prediction schemes. Manwaring et al. [7] described a portion of an experimental and computational program, which incorporates measurements of all aspects of the forced response of an airfoil row for the first time. The purpose is to extend knowledge about unsteady aerodynamics associated with a low-aspect-ratio transonic fan, where the flow defects are generated by inlet distortions. Willcox et al. [8] utilized a model order reduction technique that yields low-order models of unsteady blade row aerodynamics. The technique is applied to linearized unsteady Euler CFD solutions in such a way that the resulting blade row models can be linked to their surroundings through their boundary conditions. The technique is also applied to a transonic compressor aeroelastic analysis, which captures high-fidelity CFD forced response results better than models that use single-frequency influence coefficients.

Hall and Silkowski [9, 10] presented an analysis of the unsteady aerodynamic response of cascade due to incident gusts or blade vibration, where the cascade is part of a multistage fan, compressor, or turbine. Most current unsteady aerodynamic models assume that the cascade is isolated in an infinitely long duct. This assumption, however, neglects the potentially important influence of neighboring blade rows. Manwaring and Fleeter [11] investigated a series of experiments that is performed in an extensively instrumented axial flow research compressor to observe the physics of the fundamental flow of the unsteady aerodynamics of wake, which generated periodic rotor blade row at realistic values of the reduced frequency.

Phibel and di Mare [12] studied a comparison between a CFD and three-control-volume model for labyrinth seal flutter predictions. Peng [13] investigated a running tip clearance effect on tip vortices of induced axial compressor rotor flutter. Vasanthakumar [14] studied the computation of aerodynamic damping for flutter analysis of a transonic fan. Antona et al. [15] studied the effect of structural coupling on the flutter onset of a sector of flow-pressure turbine vanes. Srivastava et al. [16] investigated a non-linear flutter in fan stator vanes with a time-dependent fixity. Li and Wang [17] evaluated the high-order resonance of a blade under wake excitation. Johann et al. [18] investigated the experimental and numerical flutter analysis of the first-stage rotor in a four-stage high-speed compressor. McGee III and Fang [19] studied a reduced-order integrated design synthesis for a three-dimensional tailored vibration response and flutter control of high-bypass shroudless fans. Aotsuka et al. [20] focused on numerical simulation of the transonic fan flutter with a three-dimensional N-S CFD code.

Zemp et al. [21, 22] conducted an experimental investigation of the forced response of impeller blade vibration in a centrifugal compressor with variable inlet guide vanes in two parts: (1) blade damping and (2) forcing function and FSI computations. Zhou et al. [23] studied the forced response prediction for the last stage of the steam turbine blade, subject to low engine order excitation. Hohi et al. [24] investigated the influence of blade properties on the forced response of mistuned bladed disks. Siewert and Stuer [25] conducted forced response analysis of mistuned turbine bladings. Heinz et al. [26] investigated the experimental analysis of a low-pressure model turbine during forced response excitation.

Kharyton et al. [27] presented a simulation of tip timing measurements of the forced response of a cracked bladed disk. Petrov [28] studied the reduction of forced response levels for bladed disks by mistuning. Gu et al. [29] investigated the forced response of shrouded blades with an intermittent dry friction force. Green [30] presented the forced response of a large civil fan assembly. Dhandapani et al. [31] investigated the forced response and surge behavior of IP core compressors with ICE-damaged rotor blades. Lin et al. [32] simplified the modeling and parameter analysis on whirl flutter of a rotor. Tang et al. [33] conducted vibration and flutter analysis of an aircraft wing by using equivalent plate models. Zhang et al. [34] investigated the application of HHT and flutter margin method for flutter boundary prediction. Rzadkowski [35] presented the flutter of turbine rotor blades in inviscid flow. Smith [36] studied discrete sound generation frequency in axial flow turbomachinery. Lane [37] investigated system mode shapes in the flutter of compressor blade rows. Srinivasan [38] explained the flutter and resonant vibration characteristics of engine blades. Moyroud et al. [39] studied a modal coupling for fluid and structure analysis of turbomachinery flutter for application to a fan stage. Crawley [40] presented the aeroelastic formulation for tuned and mistuned rotors. Hall and Silkowski [41] and Hsu et al. [42–46] focused on the influence of neighboring blade rows on the unsteady aerodynamics of turbomachinery, flutter, and forced responses.

The unsteady analysis calculates the unsteady forcing functions of inlet distortions to calculate the forced response of turbomachinery blades. Figure 1 shows a flowchart for the flutter and forced response analysis system. This study utilizes the aeroelastic model to simulate three-dimensional aeroelastic effects by calculating the unsteady aerodynamic loads on two-dimensional strips, which are stacked from hub to tip along the span of the blade.

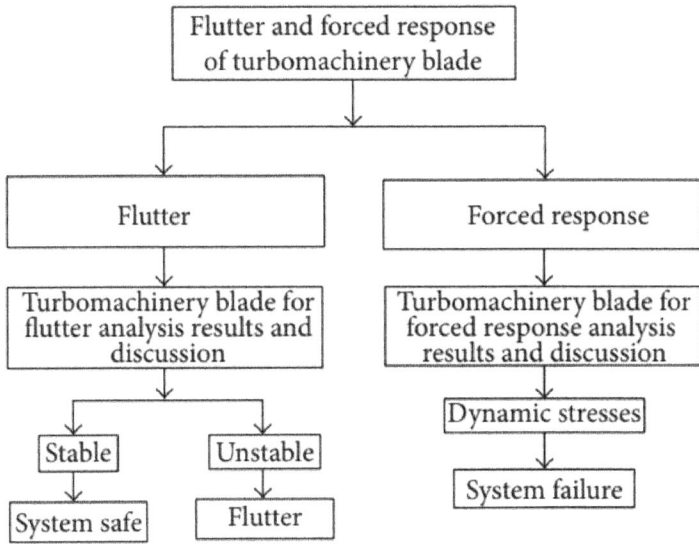

Figure 1: Flowchart for the flutter and forced response analysis system.

THEORETICAL AND NUMERICAL ANALYSIS

Analysis System

Mathematical Model

Dynamic Equation of Motion The forced response prediction system is based on an earlier developed system [11], which models the forced response of a blade caused by inlet distortion and upstream wake/shock excitation. The forced response prediction system is applied to incorporate a CFD solver to model downstream or upstream flow defects.

The forced response prediction system starts with the dynamic equations of motion, which is a system of equations for the n degrees of freedom of the system:

$$[M]\{\ddot{X}\} + [G]\{\dot{X}\} + [K]\{X\} = \{F_m(t)\} + \{F_g(t)\}.$$

(1)

The $[M]$, $[G]$, and $[K]$ matrices represent the inertia, damping, and stiffness properties of the blade, respectively, with $\{X\}$ being the n degree-of-freedom displacement. In this equation, all blades in a blade row are assumed to be vibrating as a tuned rotor, in which all blades have identical frequencies and mode shapes. The forcing terms on the right-hand side of (1) represent the motion-dependent unsteady aerodynamic forces $\{(t)\}$ and the gust response unsteady aerodynamic forces $\{F_g(t)\}$.

The solution of the undamped homogeneous form of (1) results in a set of modal properties, which are the frequencies and mode shapes for m modes. Using these modal properties, the displacements $\{X\}$ can be expressed as

$$\{X(t)\} = [\varphi]\{Q(t)\},$$

(2)

where $[\varphi]$ is the $n \times m$ mode shape matrix and $\{Q(t)\}$ is the modal displacement.

Substituting (2) with (1) and premultiplying by $[\varphi]$, the transpose of the modal matrix, results in the modal equation of motion as follows:

$$[M_m]\{\ddot{Q}\} + [G_m]\{\dot{Q}\} + [K_m]\{Q\}$$

$$= [\varphi]^T \left(\{F_m(t)\} + \{F_g(t)\} \right),$$

(3)

where

$[M_m] = [\varphi]^T[M][\varphi]$ is the generalized mass matrix,

$[K_m] = [\varphi]^T[K][\varphi]$ is the generalized stiffness matrix,

$[G_m] = [\varphi]^T[G][\varphi]$ is the generalized damping matrix,

With the assumption of simple which, in general, is a full matrix. Here, this damping matrix is assumed to be a diagonal matrix consisting of modal damping coefficients. With the assumption of simple harmonic motion, the modal displacement $\{(t)\}$ can be expressed as

$$\{Q(t)\} = \{\overline{Q}\}\, e^{i\omega t}.$$

(4)

The motion-dependent unsteady aerodynamic forces $\{(t)\}$ and the gust response unsteady aerodynamic forces $\{F_g(t)\}$ are expressed as

$$\{F_m(t)\} = [A]\{\overline{Q}\}\, e^{i\omega t},$$

$$\{F_g(t)\} = \{\overline{F}_g\}\, e^{i\omega t},$$

(5)

where $[A]$ is the unsteady aerodynamic forces due to harmonic motion of the blade and $\{\overline{F}_g(t)\}$ is the unsteady aerodynamic forces acting on the rigid blade due to a sinusoidal gust.

Substituting (4) and (5) with (3) and dividing by $e^{i\omega t}$ shows

$$-\omega^2 [M_m]\{\overline{Q}\} + i\omega [G_m]\{\overline{Q}\} + [K_m]\{\overline{Q}\}$$

$$= [\varphi]^T ([A]\{\overline{Q}\} + \{\overline{F}_g\}), \tag{6}$$

where $[A]$ is obtained by using the motion-dependent unsteady aerodynamic program with input of mode shapes and frequencies provided by a finite element vibratory analysis. $\{\overline{F}_g(t)\}$ is calculated by using the same unsteady aerodynamic program with input from a flow defect model.

Modal Aeroelastic Solution

Structural damping $[G_m]$ is estimated by using previous experience or measured data. The blade modal response is calculated with the unsteady aerodynamic loading $\{\overline{F}_g\}$, the motion-dependent unsteady aerodynamic forces $[A]$, and the structural damping $[G_m]$ as input, as seen in

$$\{\overline{Q}\} = \left[-\omega^2[M_m] + i\omega[G_m] + [K_m] - [\varphi]^T[A]\right]^{-1}[\varphi]^T\{\overline{F}_g\}. \tag{7}$$

The blade modal response $\{\overline{Q}\}$ is used to calculate the vibratory blade stress by using the modal stress information.

Model Check

A simple mode shape with only the real mode component is used to check the consistency of the complex mode flutter analysis. Two flutter analyses are performed; one with the real component mode shape $[\varphi]$ and the other with an identical mode shape, but at a different blade location of $[\varphi]^\beta$, the neighboring blade of $[\varphi]$. This identical mode shape is a complex mode shape with real and imaginary component parts. Using a single mode shape flutter analysis and a complex mode shape flutter analysis should yield the same flutter results because these two are identical mode shapes. Figure 2 shows that the two flutter analyses obtain identical results. Therefore, complex mode shapes can be used with real and imaginary mode components.

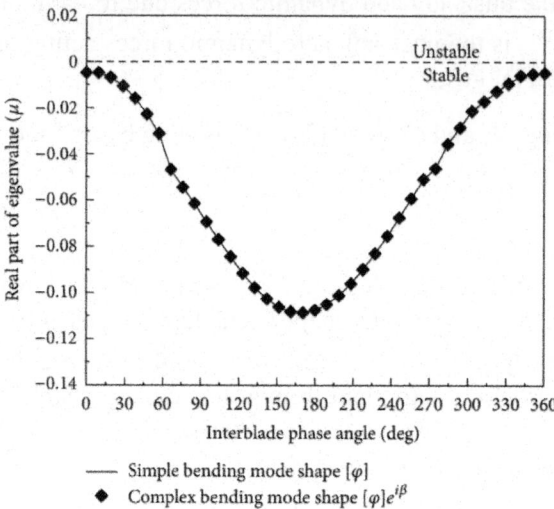

Figure 2: Complex mode flutter analysis verification.

STATIC STATE BLADE EXPERIMENTAL ANALYSIS

For the experimental testing and analysis, we used the static state blade experimental approach to measure the midspan and tip-shrouded blade response frequency and amplitude magnitude. The static state blade experimental approach uses a spectrum analyzer, a hammer for PCB model, an ICP accelerometer, a notebook/PC, rubber bands, blades, and a setup system, as shown in Figure 3.

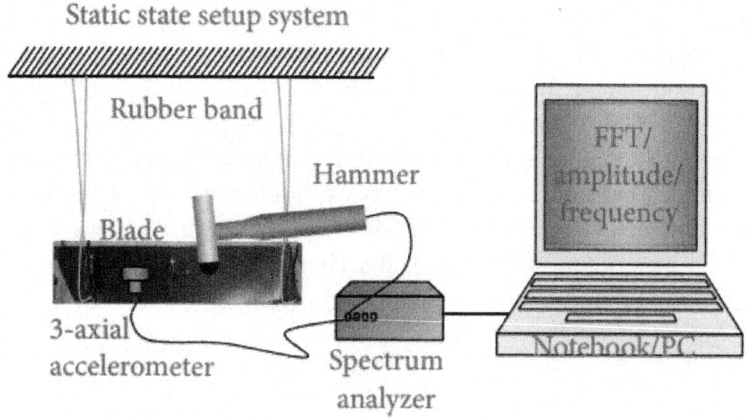

Figure 3: Static state testing and setup system.

(1) Spectrum Analyzer. PHOTON II is used to test static and dynamic signal analyses (e.g., FFT, frequency, amplitude, rpm, waterfall, dB, frequency response function, frequency response spectrum, and coherence function). According to the Nyquist rule, the measurement frequency band can be obtained 2.5 to 3.5 times, and the testing signal can be fully repeated.

  (A) Frequency Response Function. The formula for the frequency response function area is $H_1(f) = G_{xy}(f)/G_{xx}(f)$, where G_{xy} is the input and output cross frequency and G_{xx} is the power frequency.

  (B) Frequency Response The frequency response spectrum is the maximum value of the system frequency and appears as the optimal resonance value. The formula for the frequency response spectrum is $H_2(f) = G_{yy}(f)/G_{yx}(f)$, where G_{yx} is the input and output cross frequency and G_{yy} is the power frequency.

  (C) Coherence Function. Coherence Function. The formula for the coherence function area is $\gamma^2 (f) = [G(f)]^2 /(G_{xx}(f) \times G_{yy}(f)) = H_1(f)/H_2(f)$, where $0 \leq \gamma^2 (f) \leq 1$. This formula can use both the Hanning window and the exponential window.

(2) Triaxial Accelerometer (ICP number 356B21). Specifications for the triaxial accelerometer are as follows. Accelerometer sensitivity is 1.02 mV/(m/s^2) (10 mV/gn); measurement range is ±4905 m/s2 pk; frequency range is 2 Hz to 10000 Hz (y or z axis, ±5%) and 2 Hz to 7000 Hz (x axis, ±5%); resonant frequency is $\geq$55 kHz; broadband resolution (1 Hz to 10000 Hz) is 0.04 m/s^2 rms; overload limit (shock) is ±98100 m/s^2 pk; temperature range is (operating) −54∘ C to +121∘ C; excitation voltage is 18 VD to 30 VD; size is 10.2 mm× 10.2 mm × 10.2 mm; weight is 4 g; electrical connector is 8 to 36 4-pin; housing material is Ti; sensing element is ceramic; sensing geometry is shear.

(3) Hammer for PCB Model. The hammer for PCB model is used to knock the blade at different points to understand the impulse excitation material of the static state structure of the rotor blade and the natural frequency under the free-free and modal modes. The hammer is also used to knock the blade to predict the excitation frequency range of the element material, the vibration modal mode, and the physical behavior.

RESULTS AND DISCUSSION

Midspan Shrouded Blade

A midspan shrouded fan rotor is used for flutter analysis as a second option. Thirty-eight blades can be found in a fan rotor, with a midspan shroud on every blade, as seen in Figure 4. No physical connections exist between the midspan shrouds of all the blades of the common disk. However, all the shrouds make contact with one another during rotation due to the twisting of the blades.

Figure 4: Turbomachinery midspan shrouded blade model design.

Finite Element Model

The finite element model (excluding the shroud and dovetail) has 400 solid elements and 882 nodes, as seen in Figure 4. The first system mode has both bending and torsion mode components present in a single mode at the same time. The first system mode shape is decomposed into real and imaginary mode components, as shown in Figure 5.

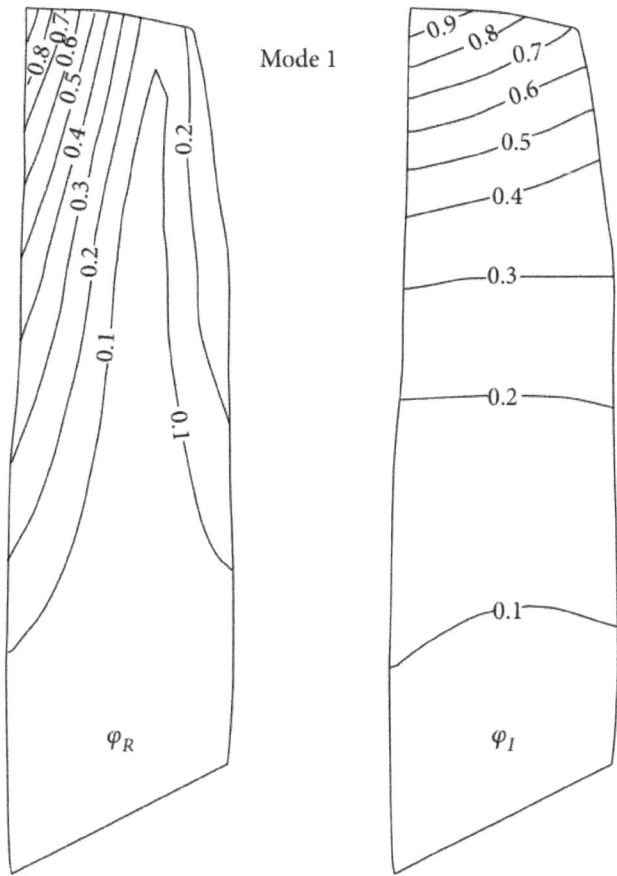

Figure 5: First system mode of midspan shrouded blade.

Flutter Stability

Figure 5 shows the decomposed real and imaginary mode components for the first system mode, where the real mode component is torsion dominated and the imaginary mode component is bending dominated. The single mode approach is used to analyze both the torsion-dominated and the bending-dominated mode shapes by assuming a similar flow field. The stability results in Figure 6 show that the torsion-dominated mode shape is unstable and the bending-dominated mode shape is stable, in which this trend is similar. With the decomposed component modes of the first system combined into the system mode as input, complex mode analysis predicted that the first system mode is stable, as shown in Figure 6. Hence, with the torsion mode component being unstable and the bending mode component being stable, the combined system mode is predicted to be stable.

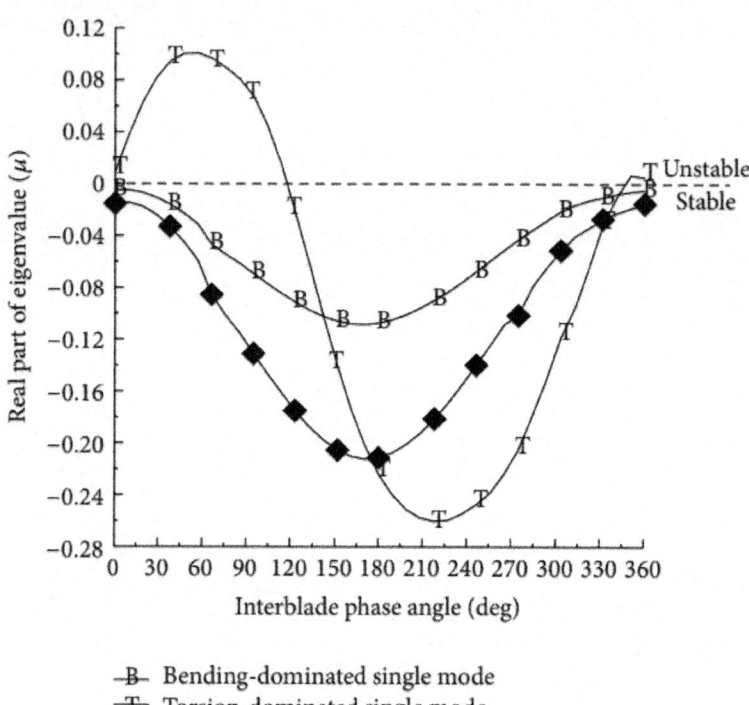

Figure 6: First system mode stability of midspan shrouded blade.

Forcing Functions

The inlet and distorted characteristics are found in six sectors, as shown in Figure 7. From the distorted characteristics, the ratio of interpolated span wise and input and computed PS, inlet total pressure, distorted static pressure, distorted tangential velocity, distorted radial velocity, total tangential velocity, axial velocity, and incidence angle characteristic results can be calculated for the six sectors, as shown in Figures 7(a) to7(h).

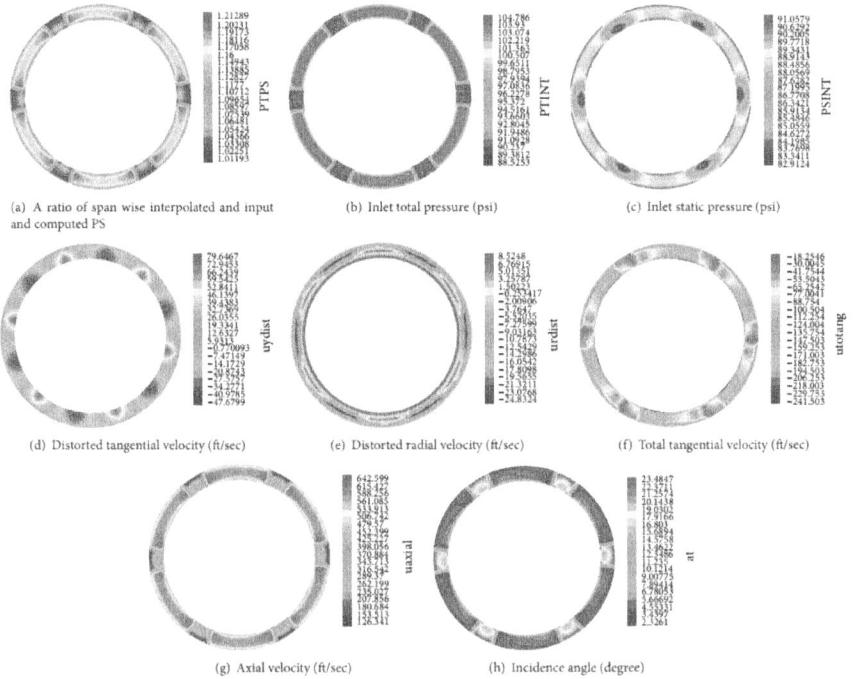

(a) A ratio of span wise interpolated and input and computed PS

(b) Inlet total pressure (psi)

(c) Inlet static pressure (psi)

(d) Distorted tangential velocity (ft/sec)

(e) Distorted radial velocity (ft/sec)

(f) Total tangential velocity (ft/sec)

(g) Axial velocity (ft/sec)

(h) Incidence angle (degree)

Figure 7: Forcing function characteristics analysis for six sectors of the midspan shrouded blade.

Turbomachinery induces six sectors of the physical characteristics of the CFD flow field interaction between air flow and the rotor blade, when the air flow inlet has six groups of midspan shrouded blades. Low and high pressures, low and high velocities, and a difference of incidence angles have inlet and distorted characteristics in the interaction area. Structural forcing bending and torsion occur in the interaction areas. The entire forcing function output database calculates the rotor complex mode and single mode forced response prediction as well as revealing the lifetime limit.

Forced Response

Figure 8 shows the Campbell diagram for the midspan shrouded rotor blade. Figure 9 shows that the single and complex modes of the midspan shrouded fan rotor blade are verified.

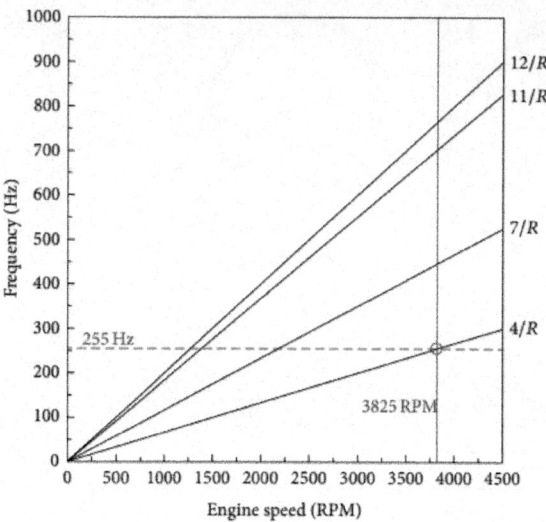

Figure 8: Campbell diagram for the midspan shrouded blade.

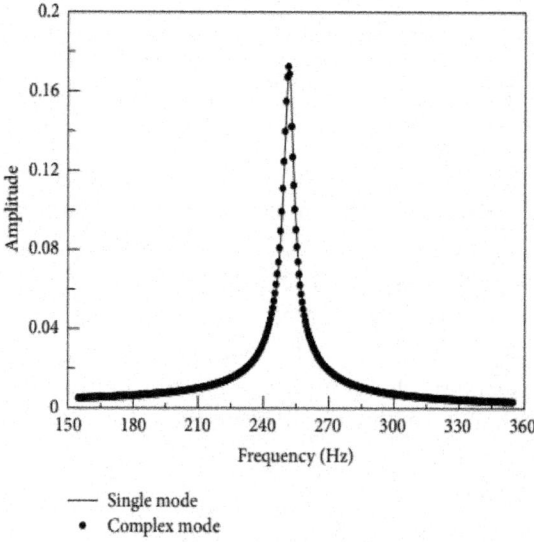

—— Single mode
• Complex mode

Figure 9: Single and complex modes verification for the midspan shrouded blade.

Figure 10 shows a phase angle of 150° for the first system mode (255 Hz) forced response analysis of the midspan shrouded fan rotor blade for the interblade. The diagram shows that the torsion mode has the highest forced response amplitude intensity, the complex mode has the lowest response intensity, and the bending mode is between the torsion and complex modes. The torsion and bending modes exceed the predicted results for the single mode approach.

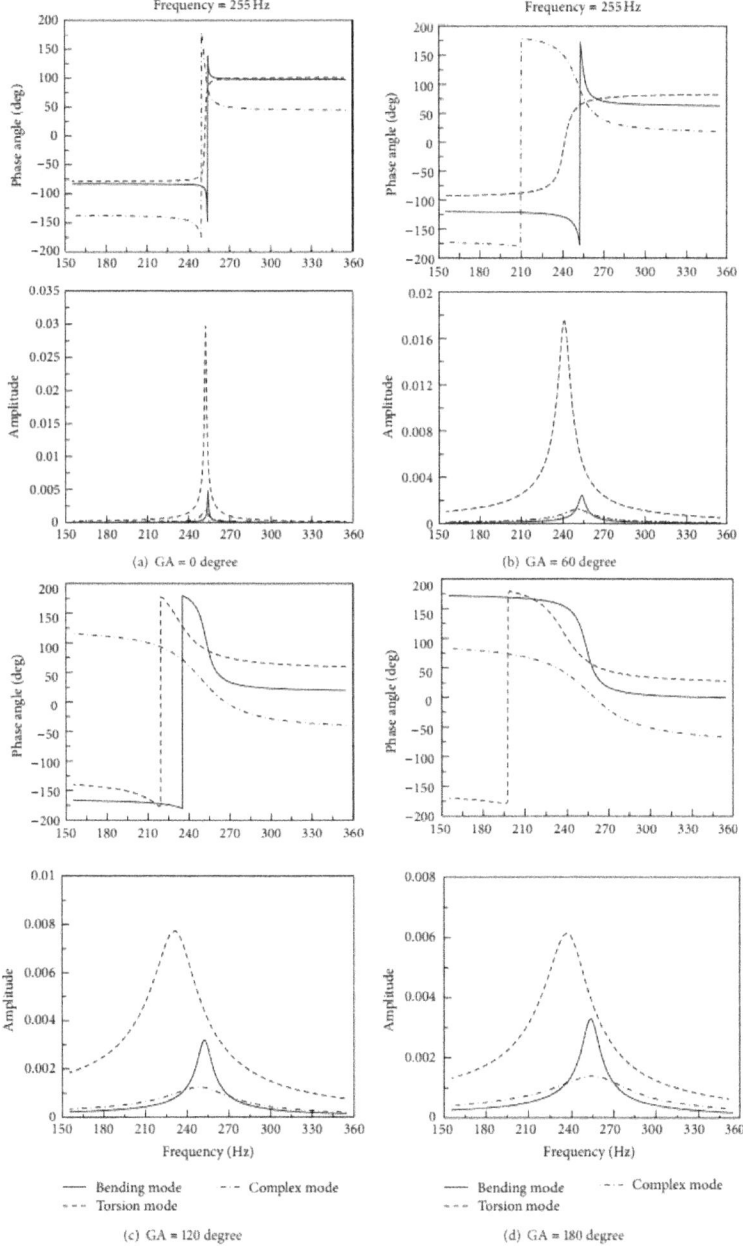

(a) GA = 0 degree

(b) GA = 60 degree

(c) GA = 120 degree

(d) GA = 180 degree

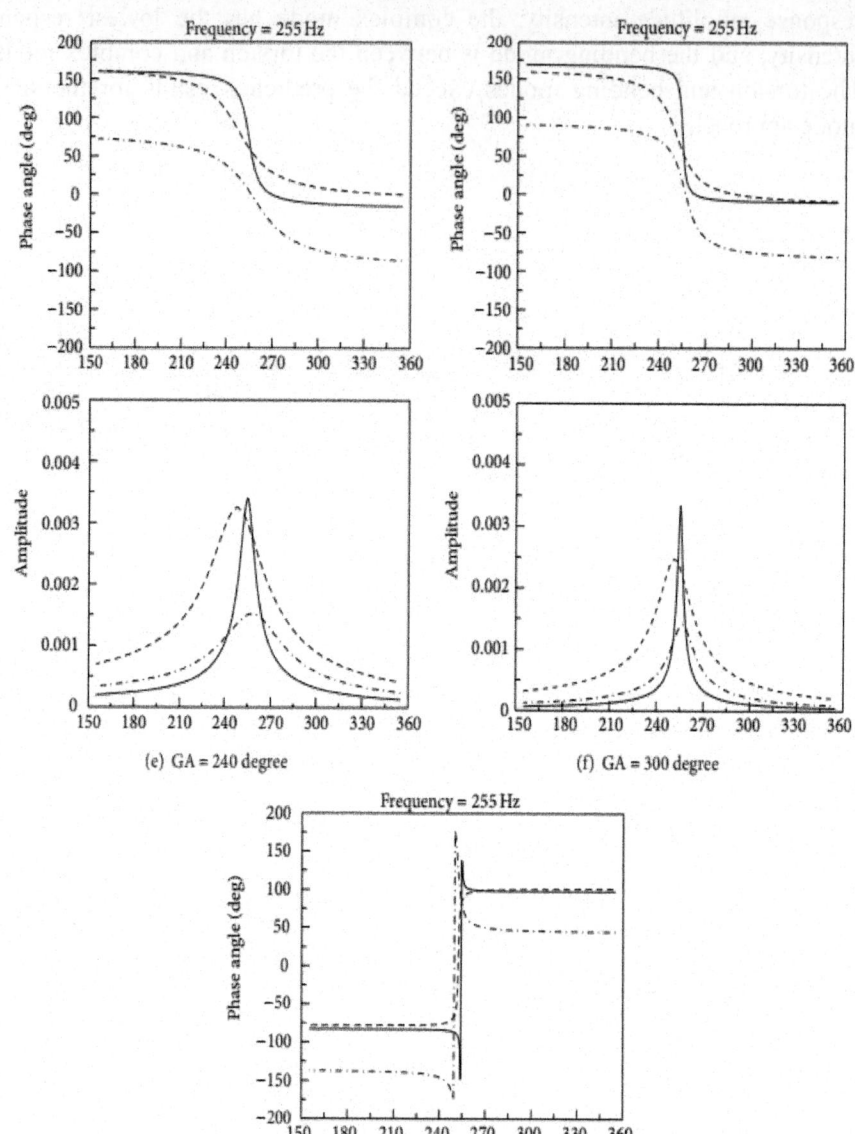

(e) GA = 240 degree

(f) GA = 300 degree

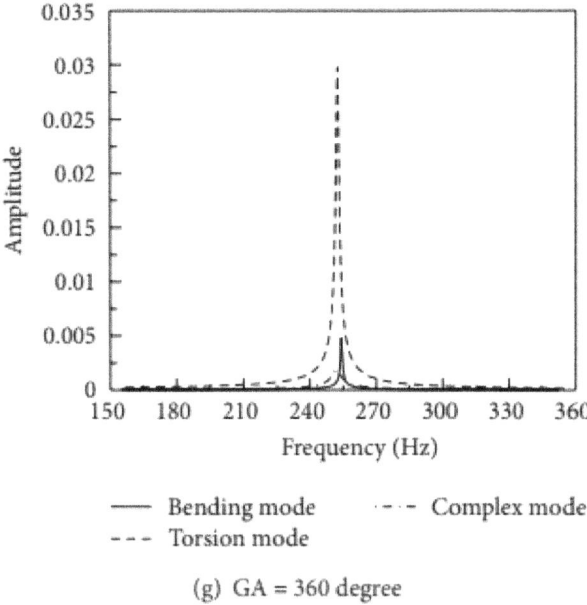

(g) GA = 360 degree

Figure 10: First system mode (255 Hz) forced response for interblade phase angles of 0, 60, 120, 180, 240, 300, and 360° of the midspan shrouded blade.

Midspan Amplitude Intensity

The amplitude intensity of the midspan shrouded blade is explained in Figure 11. The amplitude intensity and forced response increase with increasing harmonics in the tip span. The amplitude intensity and forced response increase slightly with increasing harmonics in the midspan. The amplitude and forced response decrease slightly with increasing harmonics in the hubspan. Therefore, different blade forced responses and lifetimes are induced in the different blade span areas.

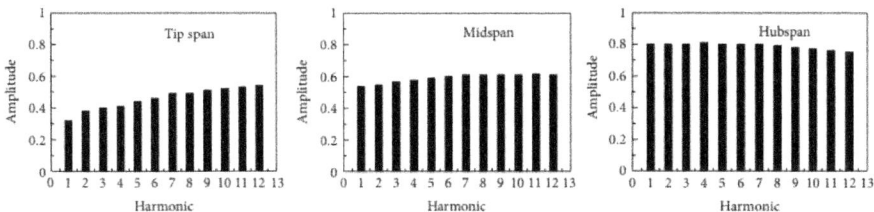

Figure 11: Amplitude intensity of the midspan shrouded blade.

Midspan Blade Static State Experiment (Similarity of Blade Testing)

The experimental frequency (Hz) and amplitude analysis (gn/LBF) of the midspan shrouded static state hammer knock blade are shown in Figures 12(a) to 12(d). Figures 12(b) to 12(d) show that the -directional main frequency is 1822 Hz and the first excitation is 539.1 Hz, the -directional main frequency and the first excitation both are 539.1 Hz, and the -directional main frequency is 1550 Hz and the first excitation is 336.9 Hz under free-free testing.

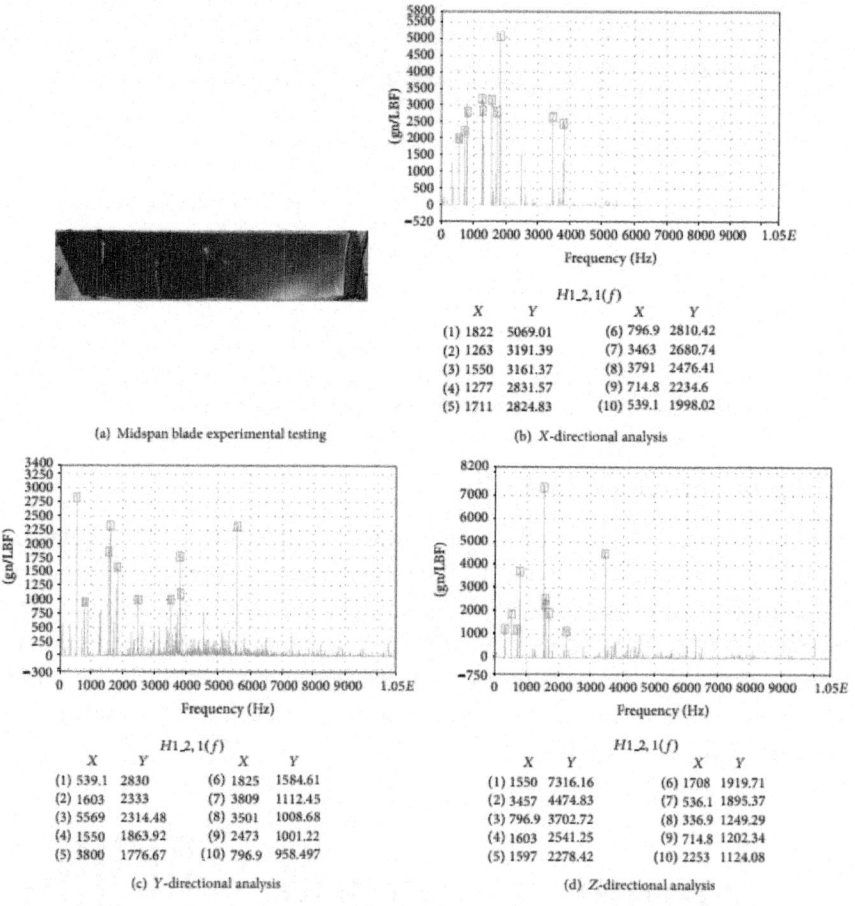

(a) Midspan blade experimental testing

(b) X-directional analysis

H1_2, 1(f)

X	Y		X	Y
(1) 1822	5069.01		(6) 796.9	2810.42
(2) 1263	3191.39		(7) 3463	2680.74
(3) 1550	3161.37		(8) 3791	2476.41
(4) 1277	2831.57		(9) 714.8	2234.6
(5) 1711	2824.83		(10) 539.1	1998.02

(c) Y-directional analysis

H1_2, 1(f)

X	Y		X	Y
(1) 539.1	2830		(6) 1825	1584.61
(2) 1603	2333		(7) 3809	1112.45
(3) 5569	2314.48		(8) 3501	1008.68
(4) 1550	1863.92		(9) 2473	1001.22
(5) 3800	1776.67		(10) 796.9	958.497

(d) Z-directional analysis

H1_2, 1(f)

X	Y		X	Y
(1) 1550	7316.16		(6) 1708	1919.71
(2) 3457	4474.83		(7) 536.1	1895.37
(3) 796.9	3702.72		(8) 336.9	1249.29
(4) 1603	2541.25		(9) 714.8	1202.34
(5) 1597	2278.42		(10) 2253	1124.08

Figure 12: Experimental analysis of the static state of the midspan shrouded blade.

Tip-Shrouded Blade

A tip-shrouded fan rotor is used for flutter analysis. Fifty blades can be found in a fan rotor, with a tip-shrouded on every blade, as shown in Figure 13. The tip-shrouded of all the blades have no physical connections, and all the shrouds are in contact during rotation due to the twisting of the blades.

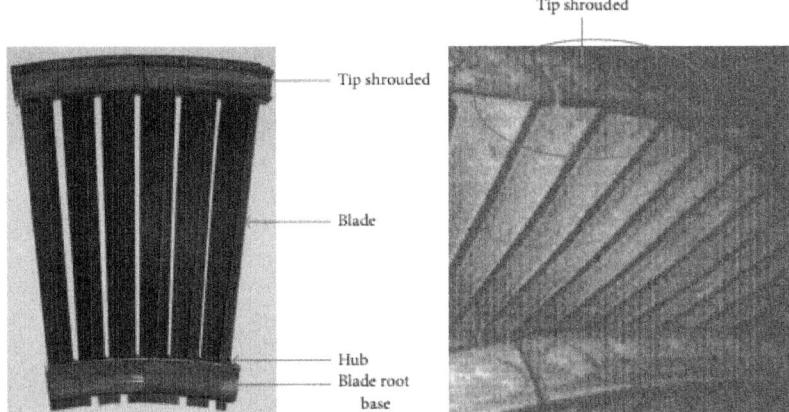

Figure 13: Tip-shrouded blade model.

Finite Element Model

As shown in Figure 13, the finite element model (excluding the shroud and dovetail) has 300 solid elements and 462 nodes. Similarly, the first system mode has bending and torsion mode components present in a single mode at the same time.

Flutter Stability

For the first system mode, Figure 14 shows the decomposed real and imaginary mode components, where the real mode component is bending dominated and the imaginary mode component is torsion dominated. The single mode approach is used to analyze both the torsion-dominated and the bending-dominated mode shapes. The stability results in Figure 15 show that the torsion-dominated mode shape is unstable and the bending-dominated mode shape is stable. With the decomposed component modes of the first system being combined into the system mode as input, the complex mode flutter analysis is used to predict the mode stability of the first system. With the torsion mode component being unstable and the bending mode component being stable, the combined system complex mode is predicted to be unstable.

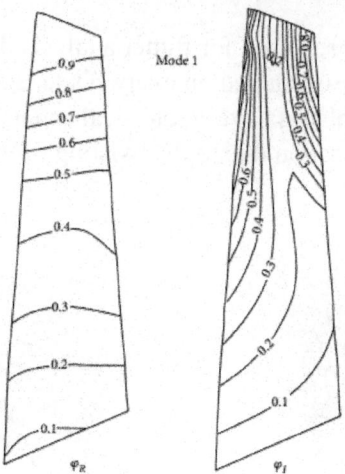

Figure 14: First system mode of the tip-shrouded blade.

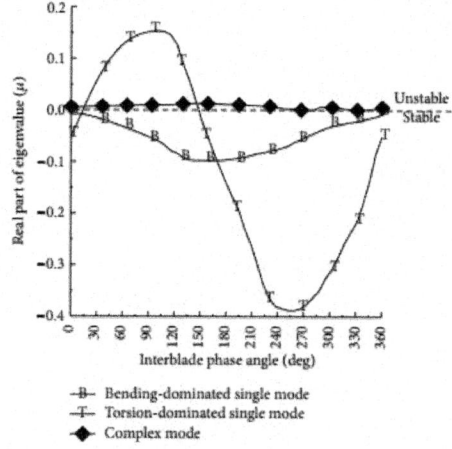

Figure 15: First system mode stability of the tip-shrouded blade.

Forcing Functions

The inlet and distorted characteristics are shown in Figure 16. From the distorted characteristics, the inlet total pressure, distorted total pressure, distorted static pressure, distorted tangential velocity, distorted radial velocity, total tangential velocity, axial velocity, summary of axial and tangential velocities, total velocity, and incidence angle characteristic results can be calculated as shown in Figures 16(a) to 16(j).

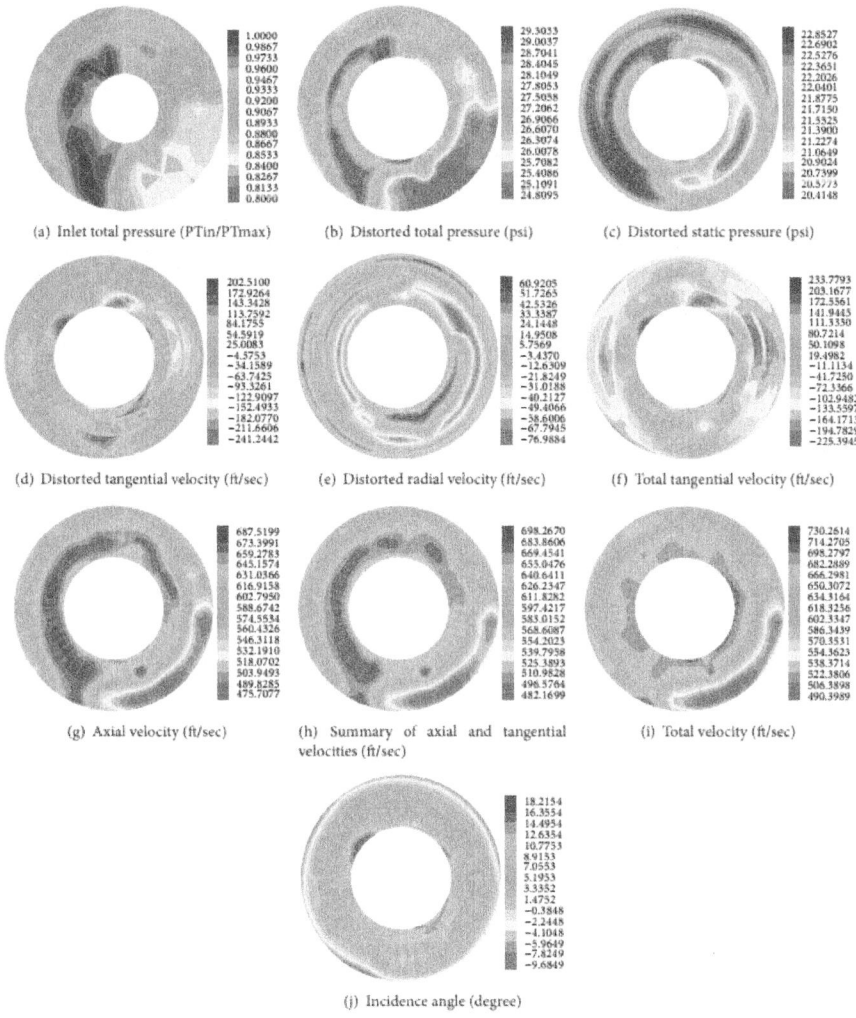

Figure 16: Forcing function characteristics analysis of the tip-shrouded blade.

Turbomachinery will induce flow field interaction characteristics between air flow and rotor blade when air flow inlet tip-shrouded blade. The inlet and distorted pressures, the difference of velocities, and the difference of incidence angles exhibit inlet and distorted characteristic distribution in the interaction area. Structural forcing bending and torsion are observed in some interaction areas. The forcing function output database can calculate forced response of the complex and single modes of the rotor as well as revealing the lifetime limit.

Forced Response

Figure 17 shows the Campbell diagram for the tip-shrouded blade. Figure 18 shows the forced response analysis of the tip-shrouded first system mode (359.1 Hz) at interblade phase angles of 0, 60, 120, 180, 240, 300, and 360°. The forced response diagram shows the complex mode, which has the highest response amplitude intensity, the bending mode, which has the lowest response intensity, and the torsion mode, which has intensity between that of the bending and complex modes. The single mode approach underpredicted the torsion and bending modes.

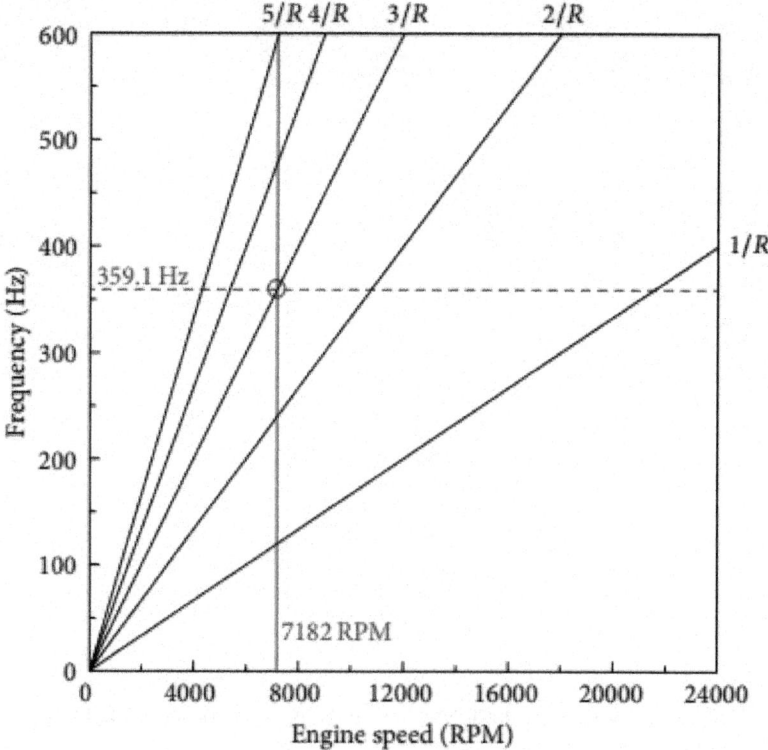

Figure 17: Campbell diagram for tip-shrouded blade.

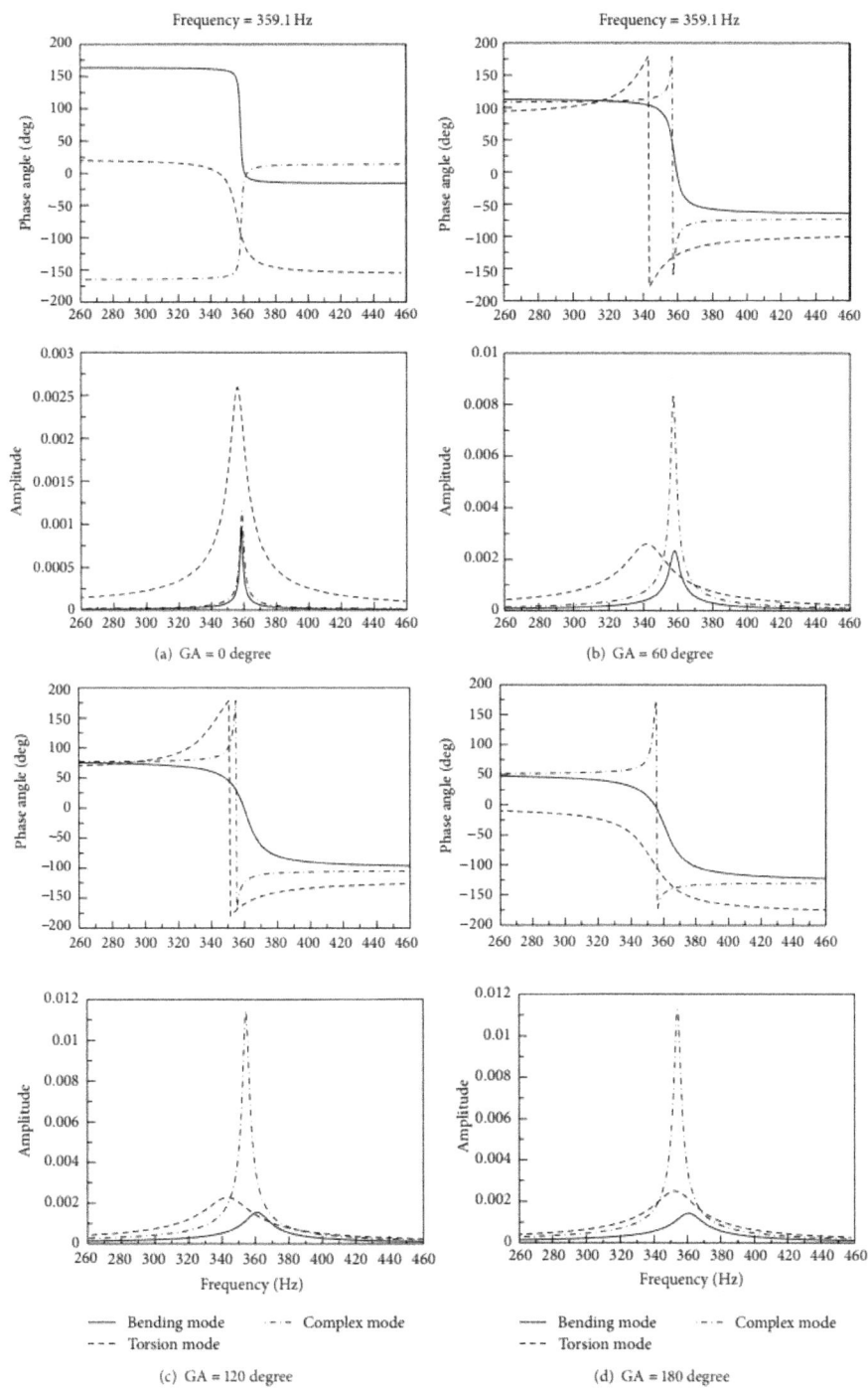

(a) GA = 0 degree

(b) GA = 60 degree

(c) GA = 120 degree

(d) GA = 180 degree

Bending mode   Complex mode
Torsion mode

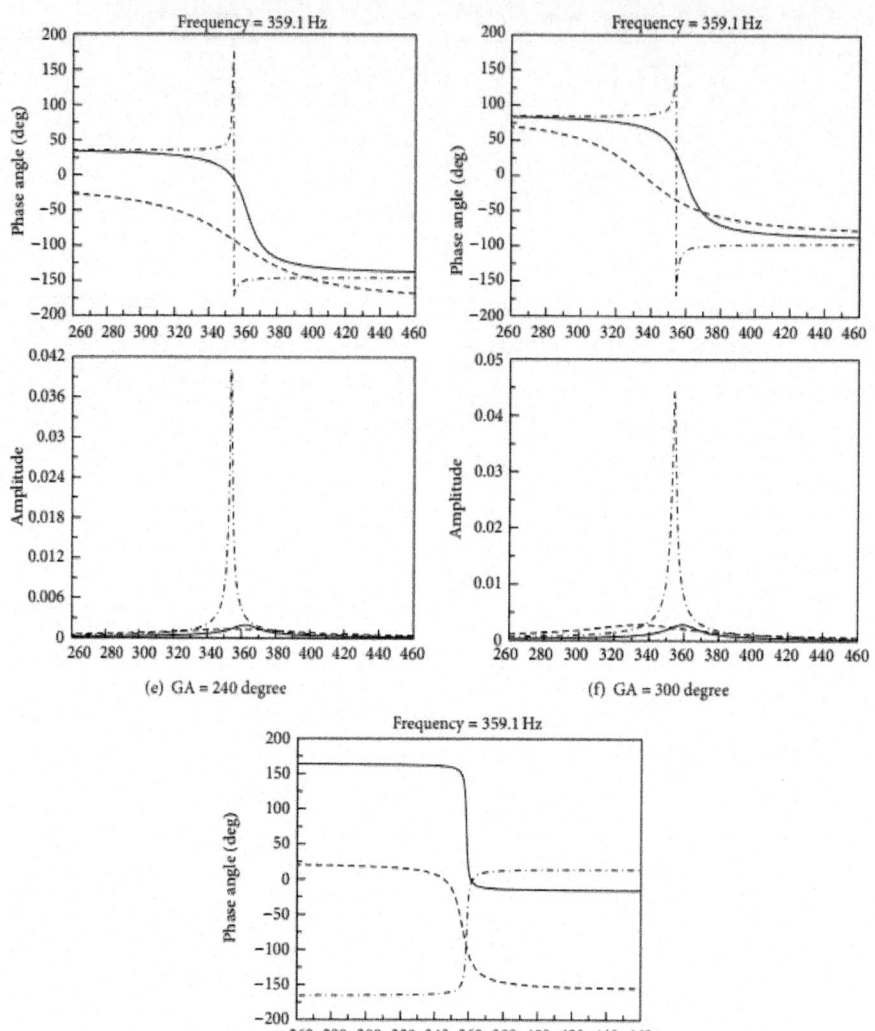

(e) GA = 240 degree   (f) GA = 300 degree

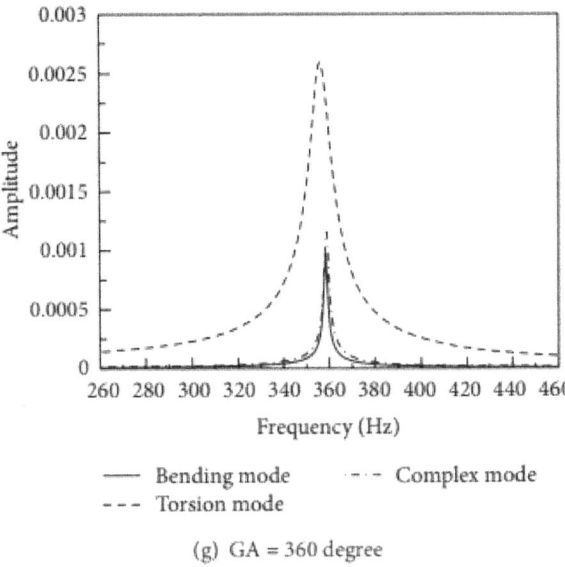

(g) GA = 360 degree

Figure 18: First system mode (359.1 Hz) forced response for interblade phase angles of 0, 60, 120, 180, 240, 300, and 360° of the tip-shrouded blade.

Tip-Shrouded Amplitude Intensity

The amplitude intensity of the tip-shrouded blade is explained in Figure 19. The amplitude intensity decreases slowly and the forced response increases and decreases slowly when the harmonic increases in the tip span. The amplitude intensity decreases rapidly and the forced response increases and decreases quickly when the harmonic increases in the midspan. The amplitude intensity and the forced response decrease and increase for one cycle when the harmonic increases in the hubspan. This condition induces different blade forced responses and lifetimes in the different blade span areas.

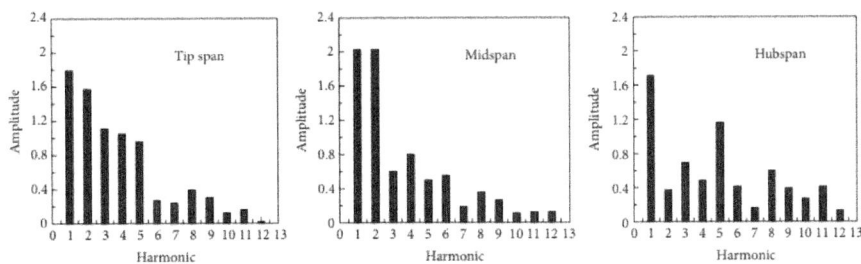

Figure 19: Amplitude intensity of tip-shrouded blade.

Tip-Shrouded Blade Static State Experiment (Similarity of Blade Testing)

The analysis of tip-shrouded blade static state experimental frequency and amplitude for the low-pressure and the high-pressure stage blades is shown in Figures 20 and 21, respectively.

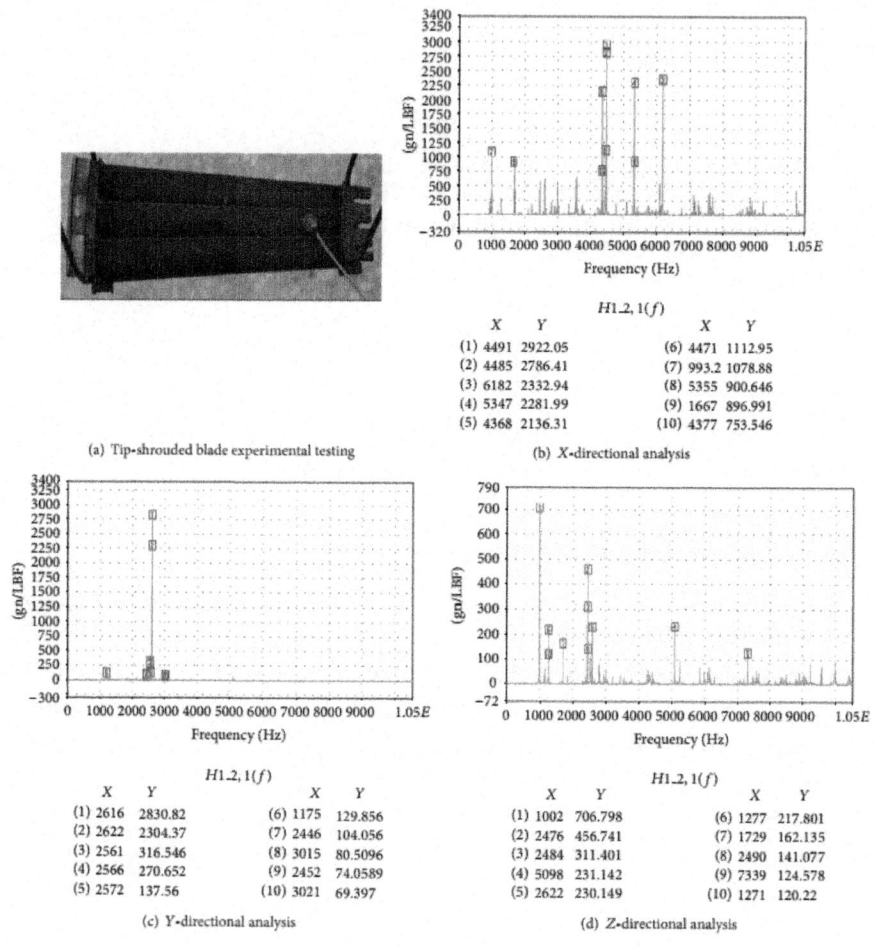

(a) Tip-shrouded blade experimental testing

(b) X-directional analysis

$H1.2, 1(f)$

	X	Y		X	Y
(1)	4491	2922.05	(6)	4471	1112.95
(2)	4485	2786.41	(7)	993.2	1078.88
(3)	6182	2332.94	(8)	5355	900.646
(4)	5347	2281.99	(9)	1667	896.991
(5)	4368	2136.31	(10)	4377	753.546

(c) Y-directional analysis

$H1.2, 1(f)$

	X	Y		X	Y
(1)	2616	2830.82	(6)	1175	129.856
(2)	2622	2304.37	(7)	2446	104.056
(3)	2561	316.546	(8)	3015	80.5096
(4)	2566	270.652	(9)	2452	74.0589
(5)	2572	137.56	(10)	3021	69.397

(d) Z-directional analysis

$H1.2, 1(f)$

	X	Y		X	Y
(1)	1002	706.798	(6)	1277	217.801
(2)	2476	456.741	(7)	1729	162.135
(3)	2484	311.401	(8)	2490	141.077
(4)	5098	231.142	(9)	7339	124.578
(5)	2622	230.149	(10)	1271	120.22

Figure 20: Experimental analysis for low-pressure stage tip-shrouded blade.

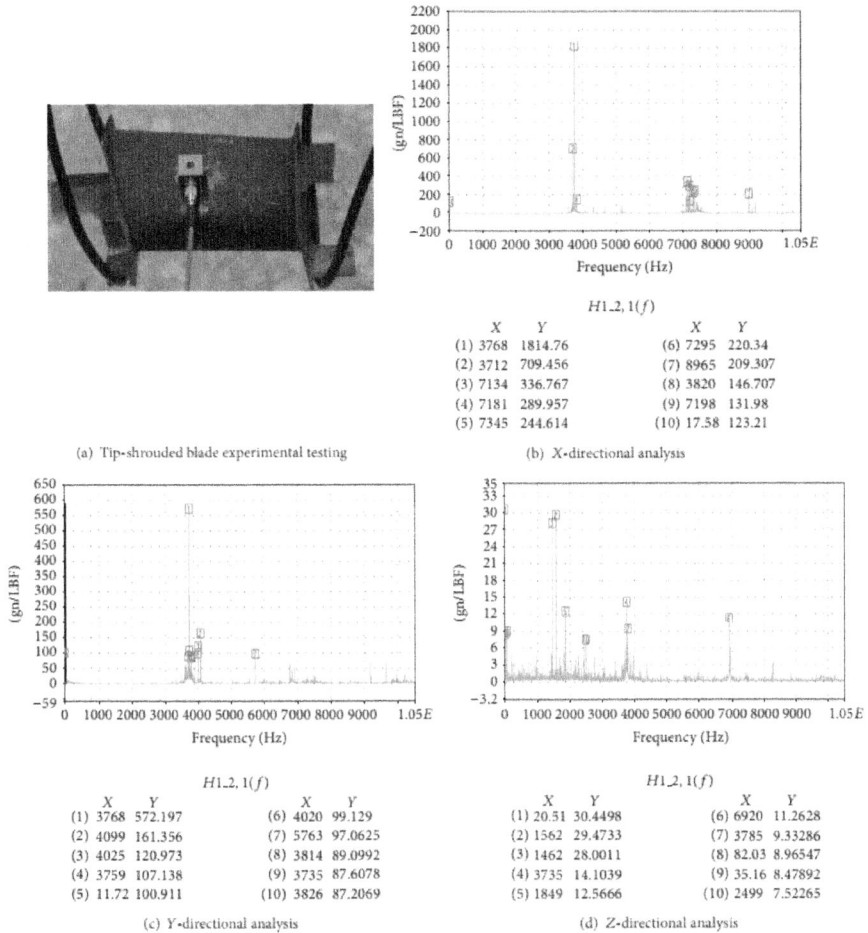

Figure 21: Blade static state experimental analysis for high-pressure stage tip-shrouded blade.

Figures 20(a) to 20(d) show the experimental frequency (Hz) and amplitude (gn/LBF) of the tip-shrouded static state knock excited blade frequency in the low-pressure stage blade. The -directional main frequency is 4491 Hz and the first excitation is 993.2 Hz; the -directional main frequency and the first excitation both are 2616 Hz; and the -directional main frequency and the first excitation both are 1002 Hz under free-free testing.

Figures 21(a) to 21(d) show the experimental frequency (Hz) and amplitude (gn/LBF) of the tip-shrouded static state knock excited blade frequency in the low-pressure stage blade. The -directional main frequency is 3768 Hz and the first excitation is 17.58 Hz; the -directional main frequency is 3768 Hz and the first excitation is 11.72 Hz, and the -directional main frequency is 20.51/1462/1562 Hz and the first excitation is 35.16/82.03 Hz under free-free testing.

CONCLUSION

Both the complex mode and single mode approaches have been individually developed to predict shrouded rotor blade flutter and forced response. A modal aeroelastic solution is also implemented. This solution models three-dimensional aeroelastic effects by calculating unsteady aerodynamic loads on two-dimensional strips, which are stacked from hub to tip along the span of the blade. A classic two-dimensional kernel function theory is utilized to calculate the unsteady subsonic, transonic, and supersonic aerodynamics, which is due to blade motion.

Finite element method is used to provide system mode shapes and frequencies. The shrouded rotor blade design causes complex blade mode shapes and, in some cases, has both bending and torsion mode components present in a single mode at the same time. Therefore, complex (cyclic symmetry) and single mode analyses are developed to predict shrouded rotor blade flutter with the combined system mode shapes of bending and torsion.

The complex and single mode analyses are useful tools for evaluating the forced response of the shrouded rotor blade, especially for cases of combined bending and torsion mode shapes. With the use of the single mode approach for the torsion and bending modes, the midspan and tip-shrouded rotor blades are over- and underpredicted by forced response analysis, respectively. With the use of the complex mode approach, the midspan and tip-shrouded rotor blades are under- and overpredicted by forced response analysis, respectively. The complex mode approach can be improved to produce the same results as the single mode approach. Therefore, the analyses conducted by using the two approaches can obtain complementary results.

ACKNOWLEDGMENT

The author would like to thank the Engineering Division of the National Science Council of the Republic of China, Taiwan, for financially supporting this research under Contract nos. NSC 100-2218-E-167-002, NSC 101-2221-E-167-013, and NSC 102-2622-E-167-022-CC3.

REFERENCES

1.   H. W. D. Chiang and R. E. Kielb, "Analysis system for blade forced response," ASME Journal of Turbomachinery, vol. 115, no. 4, pp. 762–

770, 1993.

2. H. W. D. Chiang and M. G. Turner, "Compressor blade forced response due to downstream vane-strut potential interaction," ASME Journal of Turbomachinery, vol. 118, no. 1, pp. 134–142, 1996.

3. D. V. Murthy and G. L. Stefko, "FREPS-A forced response prediction system for turbomachinery blade rows," AIAA Paper 92-3072, 1992.

4. R. E. Kielb and H. W. D. Chiang, "Recent advancements in turbomachinery forced response analyses," in Proceeding of the 30th AIAA Aerospace Science Meeting, AIAA Paper 92-0012, Reno, Nev, USA, January 1992.

5. M. S. Izsak and H. W. D. Chiang, "Turbine and compressor wake modeling for blade forced response,"ASME Paper 93-GT-148, 1993.

6. S. R. Manwaring and D. C. Wisler, "Unsteady aerodynamics and gust response in compressors and turbines,"ASME Journal of Turbomachinery, vol. 115, no. 4, pp. 724–740, 1993. ·

7. S. R. Manwaring, D. C. Rabe, C. B. Lorence, and A. R. Wadia, "Inlet distortion generated forced response of a low-aspect-ratio transonic fan," ASME Journal of Turbomachinery, vol. 119, no. 4, pp. 665–676, 1997.

8. K. Willcox, J. Peraire, and J. D. Paduano, "Application of model order reduction to compressor aeroelastic models," ASME Paper 2000-GT-0377, 2000.

9. K. C. Hall and P. D. Silkowski, "The influence of neighboring blade rows on the unsteady aerodynamic response of cascades," ASME Journal of Turbomachinery, vol. 119, no. 1, pp. 85–92, 1997.

10. P. D. Silkowski and K. C. Hall, "A coupled mode analysis of unsteady multistage flows in turbomachinery," ASME Journal of Turbomachinery, vol. 120, no. 3, pp. 410–421, 1998.

11. S. R. Manwaring and S. Fleeter, "Rotor blade unsteady aerodynamic gust response to inlet guide vane wakes," ASME Paper 91-GT-129, 1991.

12. R. Phibel and L. di Mare, "Comparison between a CFD and three-control-volume model for labyrinth seal flutter predictions," in Proceedings of the ASME TURBO EXPO: Power for Land, Sea, and Air, GT2011-46281, ASME International Gas Turbine Institute, Vancouver, Canada, June 2011.

13. C. Peng, "Tip running clearances effects on tip vortices induced axial compressor rotor flutter," inProceedings of the ASME TURBO EXPO: Power for Land, Sea, and Air, GT2011-45504, ASME International Gas Turbine Institute, Vancouver, Canada, June 2011.

14. P. Vasanthakumar, "Computation of aerodynamic damping for flutter analysis of a transonic fan," inProceedings of the ASME TURBO EXPO: Power for Land, Sea, and Air, GT2011-46597, ASME International Gas Turbine Institute, Vancouver, Canada, June 2011.

15. R. Antona, R. Corral, J. M. Gallardo, and C. Martel, "Effect of the structural coupling on the flutter onset of a sector of flow-pressure turbine vanes," in Proceedings of the ASME TURBO EXPO: Power for Land, Sea, and Air, GT2010-23037, ASME International Gas Turbine Institute, Glasgow, UK, June 2010.

16. R. Srivastava, J. Panovsky, R. Kielb, L. Virgin, and K. Ekici, "Non-linear flutter in fan stator vanes with time dependent fixity," in Proceedings of the ASME TURBO EXPO: Power for Land, Sea, and Air, GT2010-22555, ASME International Gas Turbine Institute, Glasgow, UK, June 2010.

17. L. Li and P. Wang, "Evaluation of high-order resonance of blade under wake excitation," in Proceedings of the ASME TURBO EXPO: Power for Land, Sea, and Air, GT2010-23148, ASME International Gas Turbine Institute, Glasgow, UK, June 2010.

18. E. Johann, B. Muck, and J. Nipkau, "Experimental and numerical flutter investigation of the 1st stage rotor in 4-stage high speed compressor," in Proceedings of the ASME TURBO EXPO: Power for Land, Sea, and Air, GT2008-50698, ASME International Gas Turbine Institute, Berlin, Germany, June 2008.

19. O. G. McGee III and C. Fang, "A reduced-order integrated design synthesis for the three-dimensional tailored vibration response and flutter control of high-bypass shroudless fans," in Proceedings of the ASME TURBO EXPO: Power for Land, Sea, and Air, GT2008-51479, ASME International Gas Turbine Institute, Berlin, Germany, June 2008.

20. M. Aotsuka, N. Tsuchiya, Y. Horiguchi, O. Nozaki, and K. Yamamoto, "Numerical simulation of transonic fan flutter with 3D N-S CFD code," in Proceedings of the ASME TURBO EXPO: Power for Land, Sea, and Air, GT2008-50573, ASME International Gas Turbine Institute, Berlin, Germany, June 2008.

21. Zemp, R. S. Abhari, and M. Schleer, "Experimental investigation of forced response impeller blade vibration in a centrifugal compressor with variable inlet guide vanes part 1: blade damping," inProceedings of the ASME TURBO EXPO: Power for Land, Sea, and Air, GT2011-46289, ASME International Gas Turbine, Vancouver, Canada, June 2011.

22. Zemp, R. S. Abhari, and M. Schleer, "Experimental investigation of

forced response impeller blade vibration in a centrifugal compressor with variable Inlet guide vanes part 2: forcing function and FSI computations," in Proceedings of the ASME TURBO EXPO: Power for Land, Sea, and Air, GT2011-46290, ASME International Gas Turbine Institute, Vancouver, Canada, June 2011.

23. Zhou, A. Mujezinovic, A. Coleman, W. Ning, and A. Ansari, "Forced response prediction for steam turbine last stage blade subject to low engine order excitation," in Proceedings of the ASME TURBO EXPO: Power for Land, Sea, and Air, GT2011-46856, Vancouver, Canada, June 2011.

24. Hohi, B. Kriegesmann, J. Wallaschek, and L. Panning, "The influence of blade properties on the forced response of mistuned bladed disks," in Proceedings of the ASME TURBO EXPO: Power for Land, Sea, and Air, GT2011-46826, ASME International Gas Turbine Institute, Vancouver, Canada, June 2011.

25. Siewert and H. Stuer, "Forced response analysis of mistuned turbine bladings," in Proceedings of the ASME TURBO EXPO: Power for Land, Sea, and Air, GT2010-23782, ASME International Gas Turbine Institute, Glasgow, UK, June 2010.

26. Heinz, M. Schatz, M. V. Casey, and H. Stuer, "Experimental and analysis investigations of a low pressure model turbine during forced response excitation," in Proceedings of the ASME TURBO EXPO: Power for Land, Sea, and Air, GT2010-22146, ASME International Gas Turbine Institute, Glasgow, UK, June 2010.

27. V. Kharyton, J. P. Laine, F. Thouverez, and O. Kucher, "Simulation of tip-timing measurements of a cracked bladed disk forced response," in Proceedings of the ASME TURBO EXPO: Power for Land, Sea, and Air, GT2010-22388, ASME International Gas Turbine Institute, Glasgow, UK, June 2010.

28. P. Petrov, "Reduction of forced response levels for bladed discs by mistuning: overview of the phenomenon," in Proceedings of the ASME TURBO EXPO: Power for Land, Sea, and Air, GT2010-23299, ASME International Gas Turbine Institute, Glasgow, UK, June 2010.

29. W. W. Gu, Z. Xu, and Q. Lv, "Forced response of shrouded blades with intermittent dry friction force," in Proceedings of the ASME TURBO EXPO: Power for Land, Sea, and Air, GT2008-51041, ASME International Gas Turbine Institute, Berlin, Germany, June 2008.

30. J. S. Green, "Forced response of a large civil fan assembly," in Proceedings of the ASME TURBO EXPO: Power for Land, Sea, and Air, GT2008-

50319, ASME International Gas Turbine Institute, Berlin, Germany, June 2008.

31. S. Dhandapani, M. Vahdati, and M. Imregun, "Forced response and surge behaviour of IP core-compressors with ICE-damaged rotor blades," in Proceedings of the ASME TURBO EXPO: Power for Land, Sea, and Air, GT2008-50335, ASME International Gas Turbine Institute, Berlin, Germany, June 2008.

32. L. Lin, Z. Luo, and J. Xiang, "Simplified modeling and parameters analysis on whirl flutter of tiltrotor," in Proceeding of the 8th Cross-Straits Conference on Aeronautics and Astronautics, pp. 282–289, Beijing, China, September 2012.

33. J. Tang, C. Xie, and C. Yang, "Vibration and flutter analysis of aircraft wing using equivalent-plate models," in Proceeding of the 8th Cross-straits Conference on Aeronautics and Astronautics, pp. 311–318, Beijing, China, September 2012.

34. H. Zhang, C. Xie, and C. Yang, "Application of HHT and flutter margin method for flutter boundary prediction," in Proceeding of the 8th Cross-straits Conference on Aeronautics and Astronautics, pp. 335–342, Beijing, China, September 2012.

35. R. Rzadkowski, Flutter of Turbine Rotor Blades in Inviscid Flow, AMW Wewn. 1050/2004, Akademia Marynarki Wojennej, Gdynia, Poland, 2004.

36. S. N. Smith, Discrete Frequency Sound Generation in Axial Flow Turbomachines, R & M, 3709, British Aeronautical Research Council, London, UK, 1972.

37. Lane, "System mode shapes in the flutter of compressor blade rows," Journal of the Aeronautical Science, vol. 23, no. 1, pp. 54–66, 1956.

38. V. Srinivasan, "Flutter and resonant vibration characteristics of engine blades," ASME Journal of Engineering For Gas Turbines and Power, vol. 119, no. 1, pp. 742–775, 1997.

39. Moyroud, J. G. Richardet, and T. Fransson, "A modal coupling for fluid and structure analyses of turbomachine flutter: application to a fan stage," ASME Paper 96-GT-335, 1996.

40. E. F. Crawley, "Aeroelastic formulation for tuned and mistuned rotors," in AGARD Manual on Aeroelasticity in Axial-Flow Turbomachines, vol. 2, Structural Dynamics and Aeroelasticity, 1988.

41. K. C. Hall and P. D. Silkowski, "The influence of neighboring blade rows

on the unsteady aerodynamic response of cascades," ASME Journal of Turbomachinery, vol. 119, no. 1, pp. 85–92, 1997.

42. N. Hsu, Turbomachinery blading flutter and forced response due to unsteady aerodynamics [Ph.D. Dissertation], National Tsing Hua University, 2005.

43. W. D. Chiang, C. Chen, and C. Hsu, "Prediction of shrouded turbomachinery blade flutter by a complex mode analysis," International Journal of Turbo and Jet Engines, vol. 22, no. 2, pp. 89–101, 2005. ·

44. W. D. Chiang, C. C. Chen, C. N. Hsu, G. C. Tsai, and K. L. Koai, "An investigation of turbomachinery shrouded rotor blade flutter," in Proceedings of the ASME TURBO EXPO: Power for Land, Sea, and Air, GT2003-38311, ASME International Gas Turbine Institute, Atlanta, Ga, USA, June 2003.

45. W. D. Chiang and C. N. Hsu, "Prediction of shrouded turbomachinery blade forced response by a complex mode analysis," in Proceedings of the GT ASME IGTI Turbo Expo: Power for Land, Sea and Air, GT2006-90112, Barcelona, Spain, May 2006.

46. H. W. D. Chiang and C. N. Hsu, "Turbomachine shrouded rotor blade forced response analysis,"International Journal of Turbo and Jet Engines, vol. 25, no. 3, pp. 179–188, 2008. ·

Chapter 3

CHALLENGES AND PROGRESS IN TURBOMACHINERY DESIGN SYSTEMS

R A Van den Braembussche

Von Karman Institute for Fluid Dynamics, Waterloose Steenweg, 72, B-1640 SintGenesius-Rode, Belgium

ABSTRACT

This paper first describes the requirements that a modern design system should meet, followed by a comparison between design systems based on inverse design or optimization techniques. The second part of the paper presents the way these challenges are realized in an optimization method combining an Evolutionary theory and a Metamodel. Extensions to multi-disciplinary, multi-point and multi-objective optimization are illustrated by examples.

DESIGN SYSTEM REQUIREMENTS

Navier Stokes solvers and Finite Element Stress Analysis are now routinely used to predict the performance and verify the mechanical integrity of new geometries. However they do not specify what modifications are needed to improve the performance or to minimize weight while keeping stress and vibrations below some limit values. Although they provide very detailed information, the designer will often base his decisions on overall parameters such as efficiency, pressure ratio and mass flow, leaving huge amounts of information unexploited. Present paper discusses some design systems for turbomachinery applications that have been developed over the years in order to assist the designer in finding the optimal geometry by making better use of the available information. Further progress in performance requires incorporating all 3D designs features, such as lean and sweep, that may help to improve the performance or to reach other design targets. Limitations or simplifications of the geometry are acceptable only if they are needed to satisfy other design requirements such as manufacturing or in service cost. Design systems for advanced turbomachinery should define the final geometry. Any post design geometry modification may result in a suboptimal geometry. The quality of a design depends on the accuracy of the analysis methods that have been used. Approximate solvers or surrogate models can only used for a first approximation in order to speed up the design procedures. The use of accurate solvers is mandatory to verify the final geometry because any inaccuracy of the flow solver could drive the design system towards a suboptimal geometry. The outcome of a design should not only be optimum in terms of aerodynamic/hydraulic performance but should also respect all other objectives such as cost and manufacturing limitations while assuring a safe operation over the preset lifetime. This requires a multi-disciplinary approach (fluids, stress vibration, economics etc) and a delicate balance between the different sometimes contradicting targets. Guaranteeing a stable operation over a sufficiently large operating range requires a multi-point approach. Designing for maximum efficiency or large operating range may result in different geometries or impact on the manufacturing cost. Multi-objective design systems should be able to find a compromise between these sometimes conflicting objectives. Computerized designs have the tendency to have peak performance but to be very sensitive to geometrical imperfections. All design parameters are stressed to their limit and even small variations may result in a rapid deterioration of performance. Robust design systems provide geometries that are less sensitive to geometrical imperfections and to inherent inaccuracies of the evaluation programs. An affordable computer effort is a prerequisite for a design system to be economically acceptable. Combining all previous requirements in a design system is the major challenge for the developer. Following discusses how advanced design systems respond to these challenges

INVERSE DESIGN VERSUS OPTIMIZATION

Inverse design systems define the geometry that corresponds to a pre-defined Mach number or pressure distribution. This requires a very good insight in fluid dynamics of the designer to find out how an optimum distribution should look like. This is more or less understood for 2D flows and has resulted in controlled diffusion blades. The main problem is to find out how such an optimum distribution is influenced by secondary and tip leakage flows and how to guarantee the geometrical constraints such as minimum or maximum blade thickness (existency problem). The geometry is the outcome of the inverse design so that the mechanical constraints can be verified at the end of the procedure. The latter is often avoided by specifying a thickness- and loading distribution whereby the velocity is then a consequence of the meridional contour [1]. Hence there is no direct control of the local velocity deceleration which according to Lieblein is a major factor influencing the losses. Inverse methods are often based on simplified (inviscid) flow equations eventually corrected for boundary layer blockage and neglecting secondary flows. The outcome will be different from what would be obtained by solving the real flow equations and the performance cannot be guaranteed. In what follows one will limit our self to optimization systems that are based on the accurate analysis methods commonly used now in industry. Optimization systems find the geometry that best satisfies the design objectives (OF) expressed in terms of performance, cost etc. while respecting the constraints (max. stress, lifetime). This is illustrated on figure 1 showing the iso-loss contours in function of the two design parameters X_1 and X_2.

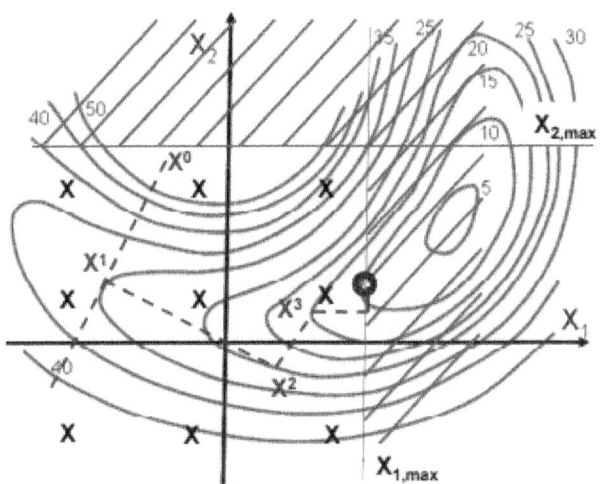

Figure 1: Optimization with two parameters

Objective is to find the combination of the design parameters X_1 and X_2 that result in a minimum loss coefficient ω while respecting the constraints. Most systems make use of an iterative procedure and start from an existing geometry (X^0). Simple mechanical and geometrical constraints ($X_1 < X_{2max}$) can already be verified before any time consuming flow analysis is started. First order methods find the optimum geometry by following the direction of steepest descend. This requires the calculation of the derivatives $_1 \partial w / \partial X$ and $_2 \partial w / \partial X$, and the calculation of the optimum step length to reach the point X_1. New steps are calculazted until the optimum geometry is found (zero gradient) or the path is blocked by a constraint. Most optimization systems make use of existing and well proven solvers to predict the OF of different geometries so that the outcome is very trustworthy. The large number of analyses that are needed to calculate the gradients and the large number of steps that may be required to reach the optimum, results in a computational effort that may be prohibitive for most real cases. A first challenge is to reduce this effort by reducing the number of required iterations and/or reducing the computational effort for each iteration. Adjoint methods allow calculating the steepest gradient with a computational burden that is comparable to the one of an analysis. This requires a modification of the flow solver, excluding the use of "off the shelf" solvers. It also complicates the extension to multisdisciplinary designs. An alternative are the zero-order or stochastic search mechanisms requiring only OF evaluations. The systematic exploration of the design space, indicated by "X" on figure 1, requires only nine OF evaluations to obtain a rather good idea of the optimum geometry. However the number of evaluations increases exponentially with the number of design parameters, leading to prohibitive computer efforts for more complex geometries. Zero order methods have fewer chances to get stuck in a local minimum. Evolutionary strategies such as Genetic Algorithms (GA), Simulated Annealing (SA), Kriging and many others can accelerate the procedure by replacing the systematic sweep by a more intelligent selection of new geometries using in a stochastic way the information obtained during previous calculations.

A way to reduce the computational burden is by working on different levels of sophistication, combining approximate but fast prediction methods with accurate but time consuming ones [2,3].

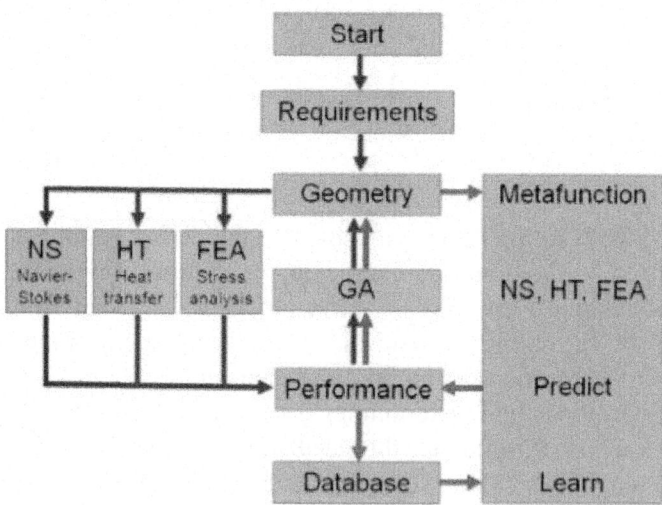

Figure 2: Flowchart of optimization system

Such a system is illustrated on figure 2. The fast but less accurate optimization loop is to the right; the expensive but accurate one is to the left. The OF driving the GA is predicted by means of a Metafunction or surrogate model i.e. an interpolator using the information contained in the Database to correlate the OF to the geometry similar to what is done by the accurate analyzers. Surrogate models have the same input and output as the analysis methods they replace. Once they have been trained on the data contained in the Database, they are very fast predictors and allow the evaluation of the OF of the many geometries, generated by the GA, with much less effort than the accurate solvers. The optimized geometry is then verified by the accurate one. The procedure is stopped when the accurate solver confirms that the surrogate model makes accurate predictions i.e. confirms that the optimizer was driven by accurate predictions. Otherwise a new GA optimization is started after a new learning of the metafunction considering also the new optimized geometries.

The main advantages of such an approach are:

- The existence of only one "master" geometry i.e. the one defined by the geometrical parameters used in the GA optimizer. This eliminates possible approximations and errors when transferring the geometry from one discipline to another.
- The possibility to shorten the design time by making all expensive analyses in parallel
- The existence of a global OF accounting for all disciplines. This allows a concurrent optimization driving the geometry to a compromise between all requirements without iterations between the aerodynamically optimum geometry and the mechanically acceptable one.

MULTIDISCIPLINARY OPTIMIZATION OF A RADIAL IMPELLER

Previous approach can easily be extended to multidisciplinary optimization by calculating the different contributions to the OF (performance, heat transfer, stress, etc.) in parallel. The method is illustrated by the design of a radial compressor impeller for a micro-gasturbine application with a diameter of 20 mm rotating at 500,000 rpm [3] The objective is a maximum efficiency while respecting the stress limits.

The first step in an optimization is the parameterization of the geometry. This is a very important issue as it should be sufficiently general, not to exclude the optimal geometry, without increasing the number of design parameters beyond a limit where it starts to slow down or prevent convergence. The 3D radial impeller is defined by the meridional contour (figure 3), in combination with the blade camber line at hub and shroud (figure 4).

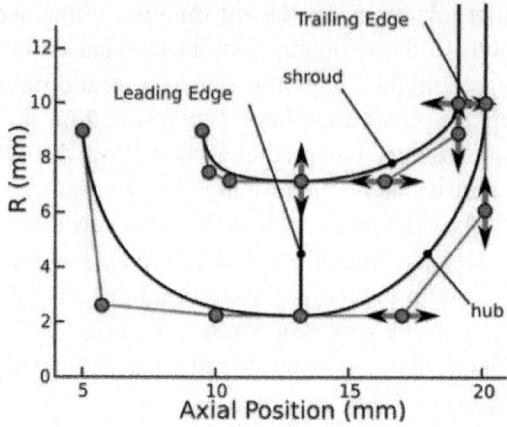

Figure 3: Meridional contour defined by Bézier control points

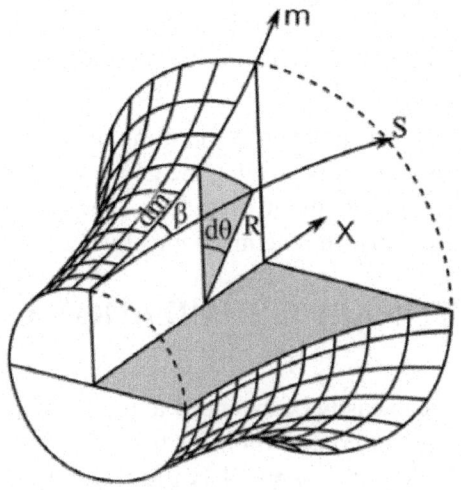

Figure 4: Definition of the blade camber line by β angle.

The meridional contours are defined by fourth order Bezier curves. Design variables are the six coordinates of the control points that can be varied and indicated by arrows on figure 3. The camberlines are defined by third order polynomials specifying the distribution of the angle β(m) between the meridional plane and the camberline (figure 4).

$$\beta(u) = \beta_0(1-u)^3 + 3\beta_1 u(1-u)^2 + 3\beta_2 u^2(1-u)^2 + \beta_3 u^3$$

u is the non-dimensional meridional length. β_0 and β_3 are the blade angles at leading- and trailing edge. The splitter blades are a short version of the full blades, with the leading edge cut back. The design is completed by a prescribed thickness distribution normal to the camber line and the number of blades. The latter could also be a design parameter to be optimized, but has been fixed to 7 for manufacturing reasons. A parameterization with more degrees of freedom is described in [4]. The OF to be minimized is a weighted sum (weight factor w) of three penalties:

$$OF(\vec{G}) = w_{stress} \cdot P_{stress}(\vec{G}) + w_{\eta} \cdot P_{\eta}(\vec{G}) + w_{massflow} \cdot P_{massflow}(\vec{G}) + w_{Mach} \cdot P_{Mach}(\vec{G})$$

The first one concerns the mechanical stresses and starts increasing when the maximum von Mises stress in the impeller σ_{max} exceeds a prefixed value $\sigma_{allowable}$.

$$P_{stress} = \max\left[\frac{\sigma_{max} - \sigma_{allowable}}{\sigma_{allowable}}, 0.0\right]$$

Expressing this constraint as a penalty does not guarantee that it is respected but has the advantage that all geometries that have been analyzed provide information that leads the GA towards the optimum geometry. The mass flow penalty increases when the error exceeds .3% of the required mass flow ($\dot{m}_{req}$ &) and with the difference in mass flow between the blade channels on both sides of the splitter blade. The latter favours an equal blade loading between splitter and full blade and improves the periodicity of the impeller exit flow

$$P_{massflow} = \left(\max\left[\left(\frac{|\dot{m}_{req} - \dot{m}|}{\dot{m}_{req}} - \frac{1}{300}\right), 0.0\right]\right)^2 + \left(\frac{\dot{m}_{upper} - \dot{m}_{lower}}{\dot{m}_{upper} + \dot{m}_{lower}}\right)^2$$

The penalty on the Mach number penalizes non-optimal loading distributions. The first part increases with negative loading. The second part increases with the loading unbalance between main blade and splitter blade. It compares the area between the suction- and pressure side Mach number distribution of main blade A_{bl} and splitter blade A_{sp}, corrected for the difference in blade length.

$$P_{Mach} = \int_0^1 \max[M_{ps}(s) - M_{ss}(s), 0.0] \cdot ds + \left(\frac{A_{bl} - A_{sp}}{A_{bl} + A_{sp}}\right)^2$$

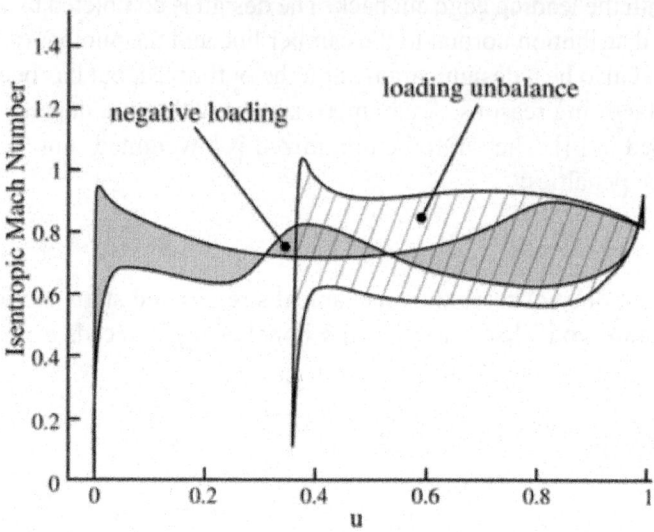

Figure 5: Negative loading and loading unbalance in a compressor with splitter vanes.

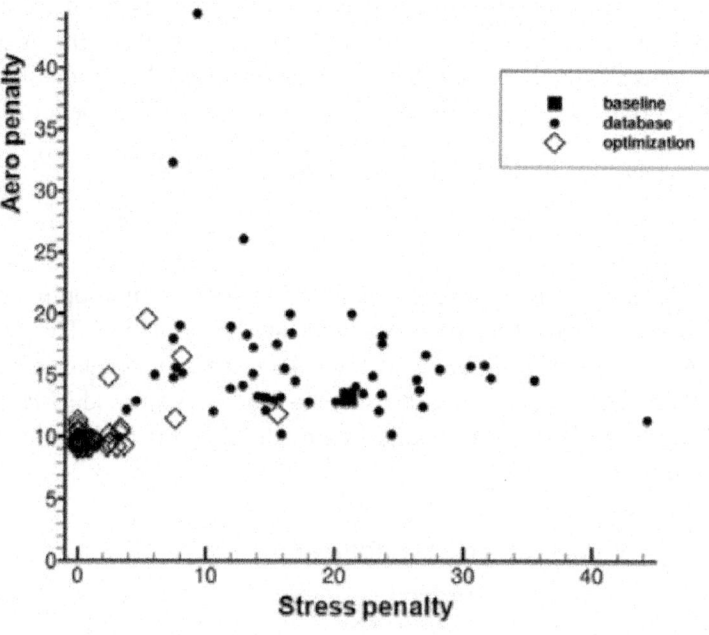

Figure 6: Aero - versus stress penalty for baseline, database- and optimization geometries.

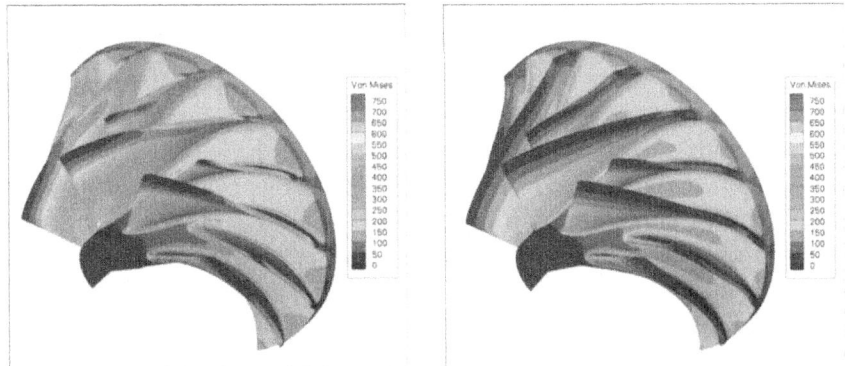

Figure 7: von Mises stresses due to centrifugal loading in the baseline (left) and optimized (right) impeller

The optimization starts from the outcome of a simple aerodynamic optimization without stress computation, called "Baseline" impeller. Although this geometry has a good efficiency, it cannot be used because a FEA stress analysis predicts von Mises stresses in excess of 750 MPa. The initial database contains 64 geometries selected by the DOE technique. About 40 optimization cycles are needed to obtain a good agreement between ANN predictions and the NS and FEA analyses. It means that the optimization is achieved with only 100 NS and FEA analyses. The aero penalty is plotted versus the stress penalty in figure 6. The geometries created during the optimization cycles are all in the region of low penalties. Most of them outperform the geometries of the Database. From all geometries at zero stress penalty, the one with minimum aero penalty is the optimum. The reduction of the maximum stress level with 370 MPa is at the cost of a 2.3 % decrease of efficiency. Figure 7 compares the von Mises stresses in the baseline geometry with the optimized one. The drastic reduction in stress is the consequence of:

- the reduced blade height at the leading edge, resulting in lower centrifugal forces at the leading edge hub
- the increase of blade thickness at the hub
- the modified blade curvature resulting in less bending by centrifugal forces

ROBUST DESIGN

Robustness expresses the in-sensitivity of the design to manufacturing noise, variations in the operating conditions and inaccuracies of the OF calculations. It is normally verified a posterior by perturbing the geometrical and operating parameters. Doing this by means of the accurate metafunction, available at the end of the design, allows very large time savings. Analysing the Database of the iterative methods, as the one presented in section 3, provides also valuable information about the robustness. This is illustrated on figure 8 showing the variation of the stress and efficiency in function of the leading edge lean. The latter is defined as the angle between the blade leading edge and the meridional plane (positive in the direction of rotation). Designers intuitively try to keep it zero in order to minimize bending stresses at the hub. Figure 8 however shows that the lowest stresses occur at -12.0°. This unexpected result is a major outcome of the optimization. The small variations of the stress and efficiency in function of lean further illustrate the robustness of this design.

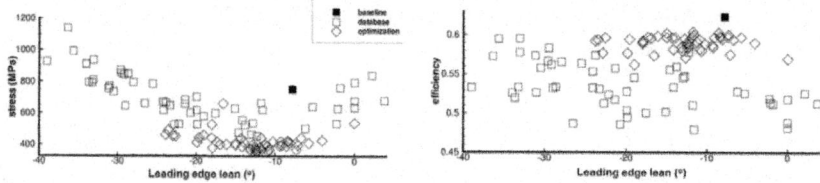

Figure 8: Blade lean versus stress and efficiency for database and optimization geometries.

MULTI-POINT OPTIMIZATION OF A TRANSONIC RADIAL IMPELLER

Multipoint designs aim to achieve sufficient range between choking and surge, to have maximum efficiency at design mass flow and a sufficiently negative slope of the pressure rise versus mass flow curve for stability reasons (figure 9). Defining the whole operating line requires a large number of expensive NS calculations that are of little use if the choking mass flow is different from the required value. The optimization algorithm has therefore been adapted by adding a geometry scaling to the optimizer which allows recuperating the optimized geometries that do not satisfy the choking requirements [4]. Geometries that show a potential in terms of operating range and performance are scaled to satisfy the choking requirements. The scaled geometry is then input to a final series of NS calculations and one FEA stress analysis. Using the results of the first series of NS analyses of the unscaled impeller one can define the boundary conditions corresponding to a uniform distribution of operating points between surge and choke. The use of databases and metafunctions dedicated to predict surge and choking further allows speeding up the convergence because fewer expensive accurate calculations will be required. The outcome of a transonic radial impeller optimization by this method is shown on figure 10. The final geometry satisfies the range requirements and the efficiency is increased by more than 2 points. The latter is due to the increased degree of freedom allowing splitter vanes that are different from the main blades. All mechanical constraints in terms of maximum stresses are respected.

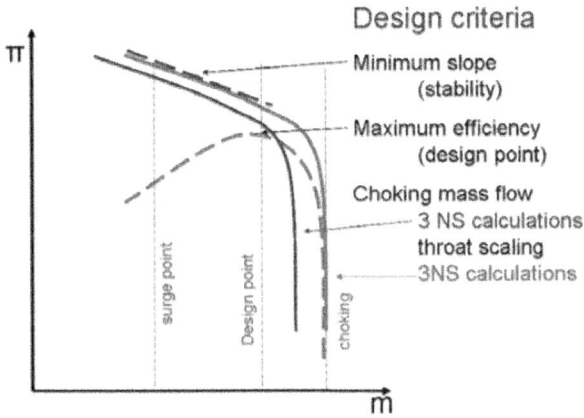

Figure 9: Procedure for multipoint compressor optimization

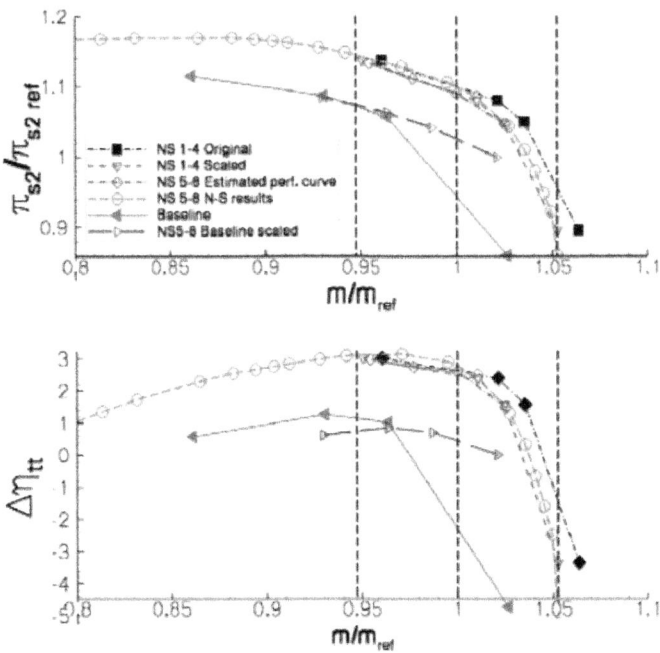

Figure 10: Performance of the redesigned versus baseline impeller

MULTI-OBJECTIVE OPTIMIZATION

The outcome of an optimization with 2 objectives can be visualized by plotting them in the 2D fitness space. The non dominated solutions i.e. the geometries G for which one objective cannot be decreased without increasing the other one, define a Pareto front. The choice is then left to the designer to select at the end of the optimization the geometry on the Pareto front that has the right balance between both objectives. A Pareto front is quite useful for problems with 2 OF as long as it remains convex. Visualization becomes more complicated when more than 2 OF are specified and special techniques are required to come to a motivated decision [5]. Defining the Pareto front is a time consuming activity requiring a large number of geometry analyses of which many will be of no interest. An alternative is a combination of the penalties corresponding to the different objectives into one pseudo-OF. Much less geometry analyses may be needed to reach the optimum.

$$OF(G) = w_1.P_1(G) + w_2.P_2(G)$$

This approach was already used in previous sections and is illustrated here by the optimization of the cooling system of a HP turbine blade [6]. The optimization aims to lower P1 (increasing with the amount of cooling mass flow) and P2 (increasing when the required life time is not reached). The lifetime depends of the equivalent stress, function of the material parameters, stress and temperature in each point of the blade. [7]. During the optimization process driven by a pseudo OF, the optimization follows a path in the design space towards the point where the lines of constant pseudo OF become tangent to the Pareto front.

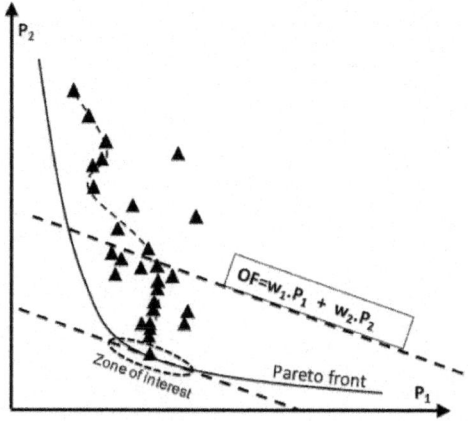

Figure 11: The evolution of P_1 and P_2 t towards the Pareto front when using a pseudo OF

The main advantage of this approach is that only 30 geometry analyses are needed to find this optimum. The disadvantage is that the pseudo OF approach requires a rather good idea of the relative weights to be given to both penalties. Increasing the weight on P1 emphasizes on minimum cooling mass flow and hence cycle efficiency. Increasing the weight on P2 emphasizes on lifetime. The choice of the relative weights is rather obvious when one objective must be satisfied without compromise. This was the case in section 4 when optimizing the radial compressor, because the stress penalty has to be satisfied at all cost. The balance between the different penalties may be less clear in other cases. However perturbing the design parameters around the optimum geometry provides information about the interesting part of the Pareto front with minimum extra effort.

CONCLUSIONS

Turbomachinery design systems based on optimization have seen a large development in recent years. The method presented here allows an important gain in design time and performance while respecting the requirements. Extensions to multipoint and multi-objective are straightforward and result in advanced and realistic geometries

REFERENCES

1. Watanabe H and Zangeneh M 2003 Design of the blade geometry of swept transonic fans by 3D inverse design ASME-GT 38770.

2. Pierret S and Van den Braembussche R A 1999 ASME Journal of Turbomachinery 121 326-332.

3. Verstraete T, Alsalihi Z and Van den Braembussche R A 2007 ASME Journal of Turbomachinery 132 03104.

4. Van den Braembussche R A, Alsalihi Z, Verstraete T, Matsuo A, Ibaraki S, Sugimoto K and Tomita I 2012 Multidisciplinary Multipoint Optimization of a Transonic Turbocharger Compressor ASME-GT 695645.

5. Sugimura K, Jeong S, Obayashi S and Kimura T 2008 Multi-Objective Robust Design Optimization and Knowledge Mining of a Centrifugal Fan that takes Dimensional Uncertainty into Account ASME-GT 51301.

6. Verstraete T, Amaral S, Van den Braembussche R A and Arts T 2008 ASME Journal of Turbomachinery 132 021014.

7. Verstraete T, Amaral S, Van den Braembussche R A and Arts T 2008 ASME Journal of Turbomachinery 132 021013.

Chapter 4

DYNAMIC BEHAVIOUR OF PUMP-TURBINE RUNNER: FROM DISK TO PROTOTYPE RUNNER

X X Huang, E Egusquiza, C Valero and A Presas

Center for Industrial Diagnostics and Fluid Dynamics (CDIF), Universitat Politècnica de Catalunya, ETSEIB Building, Av. Diagonal 647 –08028 Barcelona (Spain)

ABSTRACT

In recent decades, in order to increase output power of hydroelectric turbomachinery, the design head and the flow rate of the hydraulic turbines have been increased greatly. This has led to serious vibratory problems. The pump-turbines have to work at various operation conditions to satisfy the requirements of the power grid. However, larger hydraulic forces will result in high vibration levels on the turbines, especially, when the machines operate at off-design conditions. Due to the economic considerations, the pumpturbines are built as light as possible, which will change the dynamic response of the structures. According to industrial cases, the fatigue damage of the pump-turbine runner induced by hydraulic dynamic forces usually happens on the outer edge of the crown, which is near the leading edges of blades. To better understand the reasons for this kind of fatigue, it is extremely important to investigate the dynamic response behaviour of the hydraulic turbine, especially the runner, by experimental measurement and numerical simulation. The pumpturbine runner has a similar dynamic response behaviour of the circular disk. Therefore, in this paper the dynamic response analyses for circular disks with different dimensions and diskblades-disk structures were carried out to better understand the fundamental dynamic behaviour for the complex turbomachinery. The influences of the pattern and number of blades were discussed in detail

INTRODUCTION

In response to the oil crisis and global climate change, hydropower, as a renewable clean energy, is widely adopted in recent decades. Hydraulic pumped-storage power plants, the largest-capacity form of grid energy storage now available, can store energy in the form of water by pumping water from a lower elevation reservoir to a higher elevation, with the excess generation capacity during periods of low electrical demand. At time of peak demand, the stored water is released through turbines to follow variations of the demand. In addition to balancing the load of the electric power system, pumped-storage power plants can also used for phase modulation, frequency modulation and incident spare to improve power quality, enhance power system stability and bring a good economic benefit at the same time. The pump turbines have been well investigated and designed by the manufactories to achieve larger output power. With this tendency to raise the power concentration of the hydraulic unit, the design head, the fluid velocities and the corresponding hydraulic excitation forces have ascended remarkably. Especially, the machine, operating at the off-design conditions, will suffer high vibration levels caused by large hydraulic forces.

The dynamic responses of pump-turbines under the rising excitation forces, will affect seriously the stable operations of the hydroelectric units and the safety of power plants. For the hydraulic turbines, when the frequencies range of the pressure pulsation caused by rotorstator interaction (RSI) overlaps with same natural frequencies of the runner, and especially the vibratory mode of the excitation is the same as that of the runner, stronger vibrations and fatigue cracks will be induced on the turbines. Therefore, a good runner design should avoid the resonance. In recent years, it is not uncommon that large fatigue cracks appear on some prototype pumpturbine runners [1,2]. Plenty of papers were focused on the dynamic response of Francis runners [3-5]. Nevertheless, the reversible pump-turbine runner has a quite different design and a more complex geometry than the Francis turbine runner. In order to avoid fatigue damages on the pump-turbine runners, it is of prime importance to investigate the dynamic behavior of prototype runners. To better understand the fundamental dynamic response of these complex structures, the dynamic response analyses of geometrically similar diskblades-disk structures should be carried out. And the influences of the pattern and number of blades on the dynamic response of the runner can also be investigated.

DYNAMIC RESPONSE ANALYSIS OF PUMP-TURBINE RUNNERS

Nodal Diameter (ND) and Nodal Circle (NC)

Hydraulic turbine runner is a typical cyclically symmetric structure which is constructed by one basic sector around an axis. For such structure, it is well known that vibration mode can be characterized according to the number of its nodal diameters and nodal circles. A nodal diameter is a line that bisects the circle across the diameter and a nodal circle is a circumferential line. The mode with d ND and c NC can be denoted as mode (d, c). 2.2 Dynamic response analysis of circular disks A copper circular disk (figure 1) was selected to conduct the modal analysis. The thickness, inner and outer radius of the disk are 3mm, 13mm and 72.5mm respectively. The density of this disk is 8300 kg·m-3, the Young's Modulus is 110 GPa, and the Poisson's ratio is 0.34. The disk was treated as a free body in the simulation.

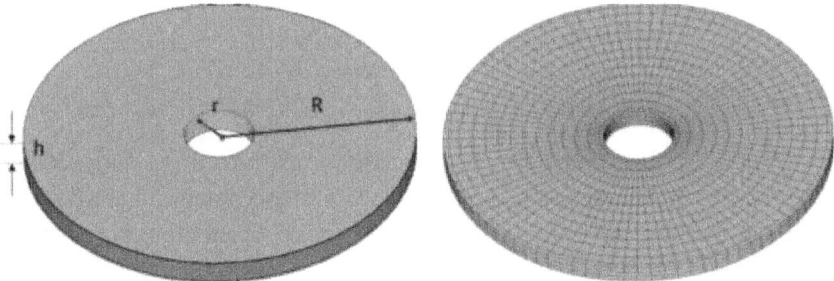

Figure 1: Geometry and finite element model of the circular disk.

The disk was suspended with flexible ropes in the measurement. The ropes have no influence on the modal behaviour of the disk, since their natural frequencies are much lower than those of the disk. The rowing hammer technique has been used for the measurements. The hammer has been moved around the structure to excite the DOFs needed to well characterize the dynamic behaviour meanwhile the accelerometers have been fixed to the structure. The numerical simulation of the circular disk can be used to identify the frequency band of interest and the regions with largest deformations for the expected mode-shapes. One piezoelectric low mass accelerometers was located on the circumference of the disk to measure the vibration response to the impulsive force excitations, and the selected 80 points were impacted by an instrumented force hammer. The signals of the impulse excitation and the response were acquired and recorded. The averaged FRFs for all the excitation positions have been calculated in order to extract the modal parameters.

Taking into account of repeated roots for the cyclically symmetric structure, the mean value of natural frequencies is adopted to describe the simulation and experiment results. Frequency deviation ratio (FDR) Δ is used to check the accuracy of the result between Sim. (RSim.) and Exp. (RExp.). The value of $\Delta(\%)$ is calculated by the following equation:

$$\Delta(\%) = 100 * (R_{Sim.} - R_{Exp.})/R_{Exp.} \tag{1}$$

The natural frequencies and FDR are shown in table 1. The FDR values of all the obtained modeshapes remain in a range of $\pm 2.5\%$.

Table 1: The natural frequencies and FDR of the disk

Mode-shape	Sim.	Exp.	$\Delta(\%)$
(2,0)	1031.55	1050	-1.76
(0,0)	1747.60	1790	-2.37
(3,0)	2434.15	2440	-0.24
(1,1)	4012.05	3985	0.68
(4,0)	4252.30	4210	1.00
(5,0)	6481.85	6360	1.92
(2,1)	6796.90	6780	0.25
(0,1)	7671.20	7730	-0.76

Figure 2 illustrates the mode-shapes estimated by simulation and experiment, which serve to characterize the vibration modes. For all the modes, the experimental and numerical results agree with each other very well, which validates the simulation method.

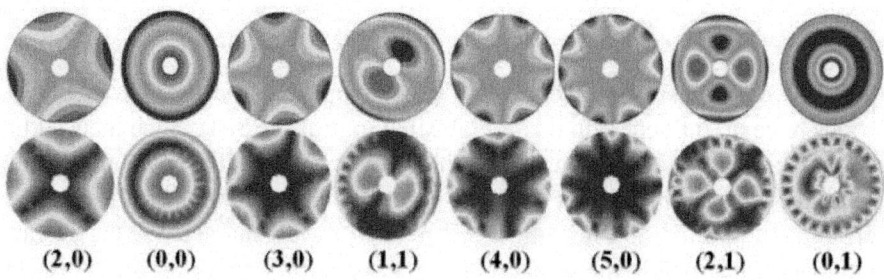

(2,0)   (0,0)   (3,0)   (1,1)   (4,0)   (5,0)   (2,1)   (0,1)

Figure 2: Mode-shapes of the disk obtained by Sim. and Exp.

With the same material parameters, dynamic response analyses of 9 disks with different radius and thickness (table 2) was performed. The natural frequencies for the corresponding mode-shape of the disks are shown in figure 3.

Table 2: The dimensions of the disks

(mm)	No.1	No.2	No.3	No.4	**No.5**	No.6	No.7	No.8	No.9
h	3	6	12	3	**6**	12	3	6	12
r	6.5	6.5	6.5	13	**13**	13	26	26	26
R	36.25	36.25	36.25	72.5	**72.5**	72.5	26	26	26

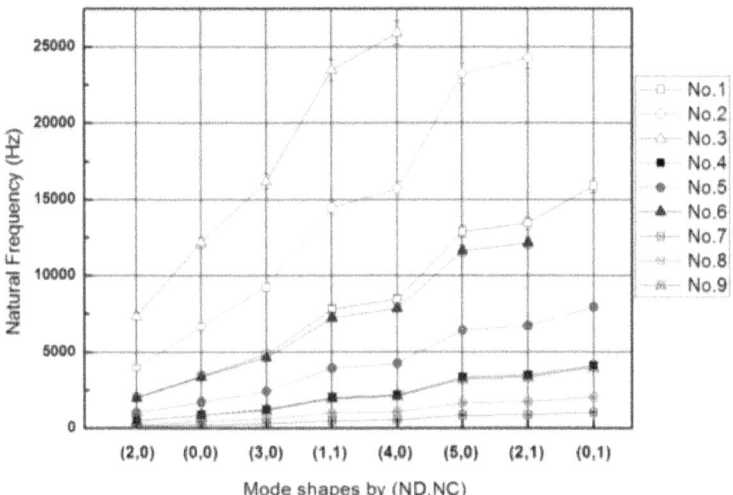

Figure 3: Natural frequencies of the investigated disks.

The corresponding mode-shapes of different disks are similar with large deformation in radial direction. For any specified mode-shape, the natural frequencies for the disks with the same radius dimensions increase with the increasing thickness, and the corresponding frequencies increase almost twice with the doubled thickness. The natural frequencies for the disks with the same thickness decrease with the increasing radius dimensions, and the corresponding frequencies are reduced to a quarter of the former one when the radius dimension is doubled. 2.3 Dynamic response analysis of model and prototype pump-turbine runner With the same procedure, the dynamic response analysis of the scaled model runner and prototype runner were carried out by experiment and simulation. The obtained mode-shapes are illustrated in figure 4. The FDR and the corresponding natural frequencies of modal runner obtained from simulation and experiment are listed in table 3. The experimental and numerical results have a rather good agreement. The first several modes of the model and prototype runners give the same modeshapes as the disks.

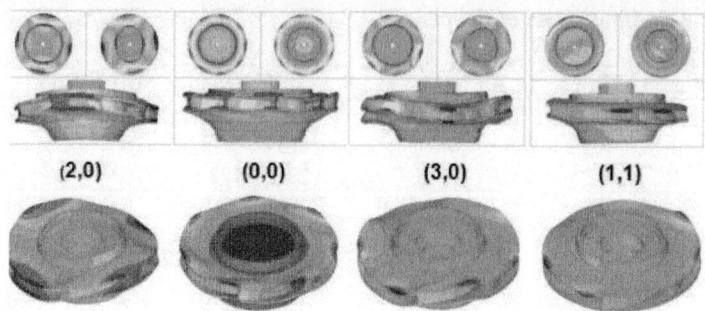

Figure 4: Mode-shapes of the modal runner(Up) and the prototype runner (down)

Table 3: Natural frequencies and FDR of the modal runner in air.[6-9]

Mode-shape	Model runner			Prototype runner		
	Exp.	Sim.	Δ(%)	Exp.	Sim.	Δ(%)
(2,0)	825.26	838.00	-1.52	250.93	250.41	-0.21
(0,0)	1247.40	1111.00	12.28	320.96	337.04	5.01
(3,0)	1479.75	1463.00	1.14	435.65	439.89	0.97
(1,1)	1605.55	1439.00	11.57	424.30	431.81	1.77

Dynamic Response Analysis Disk-Blades-Disk Structures

In order to gain a better understanding of the complicated dynamic response of pump-turbine runners, it is better to investigate the corresponding simplified models. Since the pump-turbine runner is consist of three parts: crown, blades and band, it can be simplified as disk-blades-disk structure. Dynamic response analyses of disk-blades-disk structures were performed to understand the influences of the basic dimensions (such as shape and number of blades) on the dynamic response the structures. The investigated disk-blades-disk structures consist of a couple of circular disks and several curved or straight blades (figure 5). The circular disks have the same dimensions as the previously described disk No.5. The height and width of the blades are 13.5mm and 3mm.

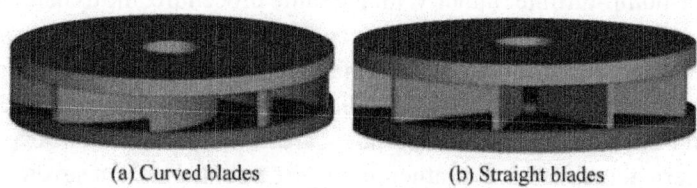

(a) Curved blades   (b) Straight blades

Figure 5: Geometries of disk-blades-disk structures.

The modal analysis of the disk-blades-disk structures with 7-9 curved and straight blades was carried out. The natural frequencies of the same mode-shapes are listed in table 4.

Table 4: Natural frequencies of the disk-blades-disk structures

Mode-shape	Natural frequencies							
	Curved blades				Straight blades			
	6 blades	7 blades	8 blades	9 blades	6 blades	7 blades	8 blades	9 blades
(2,0)	2633.70	2754.60	2850.05	2931.40	2223.70	2330.50	2409.30	2471.90
(0,0)	3289.50	3533.70	3740.80	3933.90	4790.90	5226.50	5542.10	5776.30
(3,0)	4542.80	4668.90	4939.15	5151.30	3698.00	3893.50	4132.00	4304.50
(1,1)	5032.60	5235.50	5414.75	5578.40	4624.00	4830.10	5010.30	5173.50

Since the corresponding mode-shapes of the structures with different number of blades are similar, the mode-shape of the disk-blades-disk with 9 curved and straight blades are shown in figure 6.

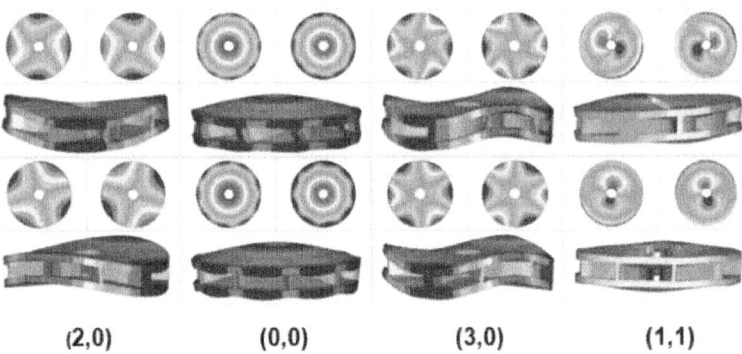

(2,0)   (0,0)   (3,0)   (1,1)

Figure 6: Mode-shapes of the disk-blades-disk structures with curved blades (up) and straight blades (down).

Although the number and the shape of the blades are changed, the disk-blades-disk systems give the same mode-shapes. The mode-shapes like (2,0), (0,0), (3,0) and (1,1) can be found for the structures with curved and straight blades, but the mode-shape (0,0) for the structure with straight blades swaps the positions with (3,0) and (1,1). Various number and shape of the blades will modify the mass and stiffness of the structure, therefore the corresponding natural frequencies of the disk-blades-disk structures will be changed. For the same mode-shape, the more blades the structure has, the higher natural frequency is. For modeshapes (2,0), (3,0), (1,1), the structure with straight blades, compared with the one with the same number of curved blades, has lower natural frequencies. It is opposite for the mode-shape (0,0).

CONCLUSIONS

The dynamic response behaviours of the structures, from disk to prototype type pump-turbine runner, have been analyzed numerically and experimentally. The comparison between the experiment and simulation gives a good agreement. The model and prototype runners give the same mode-shape as the disks and disk-blades-disk structures. The vibration modes represent the typical mode-shapes for the cyclic geometry (2,0), (0,0), (3,0), (1,1) etc. For the specified mode-shape of the disks, the natural frequencies increase with the increasing thickness and decreasing radius. The number and shape of the blades will not change the mode-shapes of the disk-blades-disk structures but change the natural frequencies. For the specified mode-shape, the structure with more blades has a higher natural frequency. The shape of blades can increase or decrease the natural frequencies of the structure sdepending on the mode-shapes. So the designer can change the number or the shape of the blades to modify the natural frequencies and/or the exciting mode of the runner. From the industrial point of view, the modal parameters obtained from these simplified models can be adopted to speed-up the design process and build-up a realistic turbine with enough accuracy to decrease the vibration levels and avoid resonance and fatigue problems.

REFERENCES

1.   Egusquiza E, Valero C, Huang X et al 2012 Engineering Failure Analysis 23 27-34.

2.   Huang X 2011 Contribution on the Dynamic Response of Large Hydraulic Turbomachinery Components (Barcelona: Universitat Politècnica de Catalunya).

3.   Rodriguez C, Egusquiza E, Escaler X, Liang Q W and Avellan F 2006 J Fluids Struct. 22(5) 699-712.

4.   Liang Q W, Rodriguez C G, Egusquiza E, Escaler X and Avellan F 2007 J Computers & Fluids 36 1106-18.

5.   Lais S, Liang Q W, et al 2009 International Journal of Fluid Machinery and Systems 2(4) 303- 314.

6.   Escaler X et al 2010 Modal behavior of a reduced scale pump-turbine impeller. Part I. Experiments 25th IAHR Symposium on Hydraulic Machinery and Systems (Timişoara, Romania, 20-24 September 2010) 918-926.

7.   Valero C, Huang X, et al 2010 Modal behavior of a reduced scale pump-turbine impeller. Part II. Numerical Simulation 25th IAHR Symposium

on Hydraulic Machinery and Systems (Timişoara, Romania, 20-24 September 2010) 926-934.

8. Egusquiza E, Valero M and Huang X 2009 The effect on boundary condition on the structural response of a pump-turbines 3rd IAHRWG Meeting (Brno, Czech Republic, 14-16 October 2009).

9. Rodriquez C 2006 Feasibility of on board measurements for Predictive Maintenance in Large Hydraulic Turbomachinery (Barcelona: Universitat Politècnica de Catalunya).

Chapter 5

EMPIRICAL DESIGN CONSIDERATIONS FOR INDUSTRIAL CENTRIFUGAL COMPRESSORS

Cheng Xu and Ryoichi S. Amano

Department of Mechanical Engineering, University of Wisconsin-Milwaukee, Milwaukee, WI 53212, USA

ABSTRACT

Computational Fluid Dynamics (CFD) has been extensively used in centrifugal compressor design. CFD provides further optimisation opportunities for the compressor design rather than designing the centrifugal compressor. The experience-based design process still plays an important role for new compressor developments. The wide variety of design subjects represents a very complex design world for centrifugal compressor designers. Therefore, some basic information for centrifugal design is still very important. The impeller is the key part of the centrifugal stage. Designing a highly efficiency impeller with a wide operation range can ensure overall stage design success. This paper provides some empirical information for designing industrial centrifugal compressors with a focus on the impeller. A ported shroud compressor basic design guideline is also discussed for improving the compressor range.

INTRODUCTION

New compressor designs always must meet the customers' needs with the shortest time to market, low cost, and improved performance. To push the design to state of the art aerodynamic performance, the structure design also needs to meet a suitable performance life of the compressors. Mechanical integrity is one of the important parts of the centrifugal compressor design. Mechanical constraints are usually negative factors for aerodynamic design, for example, mechanical constraints require thick blade for reliability but hurt impeller efficiency. The purposes of the mechanical analyses are to provide all compressor components within a reasonable time duration to sustain the aerodynamic and centrifugal force, and eigen frequencies do not match critical excitation frequencies [1]. The safety factors of the mechanical design had been reduced dramatically compared with "old fashioned" design. Due to the nature of the Finite Element Analysis (FEA) tools and material property improvements, the safety factor of a modern industrial compressor design normally is set to 7 to 12%. The mechanical requirements need structure designers to have better practice to allow more freedom to aerodynamic designers and to keep all the components at the lowest weight and the lowest cost.

Design of a long lifetime single component of compressors is not a goal for designers. Emphasis on improving efficiency has been a primary issue, but this also is not as important as in the past. The development cost and development time is also a key factor that needs to be considered for a modern compressor design. Industrial compressor design expects a state-of-the-art performance compressor without making a second build for less cost and short development time. For achieving this goal, compressor design engineers need to have multidiscipline knowledge of centrifugal compressor design. Detailed design considerations can reduce the time to perform the advance design studies and laboratory investigations. The wide variety of design subjects represents a very complex design world for compressor designers. One purpose of this paper is to provide information in an aerodynamic point of view to understand the overall design before starting the detailed design process of a centrifugal compressor. The paper also summarizes important aspects of the centrifugal compressor design for industrial compressor designers and scientists.

The compressor market and business model has changed in the last few decades. Industrial compressor design now requires designing for success in the marketplace, not just for scientific experiments. In the past, compressor designers developed a new compressor in the development group and passed the design to manufacturing. The manufacturing group would evaluate how to make it at the lowest cost, and some designs were rejected because they could not meet the market requirements. The new development model requires the compressor designer to design for market, manufacturing, and end users. New business concepts have been proposed [2, 3] in which the design also considers an integrated system of manufacturers and end users. The new compressor developments become a complex system task. Minimizing manufacturing cost of the compressor design is not enough. The compressor design must consider all aspects of the manufacturing and end users. If surplus is defined as the total profit of manufacturing, end users, and aftermarket, the compressor's new development will focus on the design for maximum surplus. Therefore, in the compressor design stage, many choices of design options need to be considered before the final design, and discussions must consider the surplus value. It is essential that design engineers begin to perform a compressor design with full understanding of all aspects of the design considerations [4–8].

To reduce manufacturing cost, many high volume compressor manufacturers, for example turbocharger, often use the flow cut for different applications. The flow cut uses the same defined blade geometry for multiple flows and a similar or lower pressure ratio. This is different from scaling, in that the impeller blade and any diffuser vane geometry maintain the same definition. A brief introduction of the flow cut is also discussed in this paper.

With the development of computational science and computer hardware, design engineers rely on quality models to establish the physical relationships among diverse thermodynamic, geometric, and fluid dynamic parameters that govern turbomachinery performance [9–26]. Although CFD has helped to design many successful industrial compressors and has become an important tool in industrial compressor design, multidomain optimisation is still very time consuming. Most CFD optimisations still focus on the component [22]. When the compressor inlet flow is reduced, the compressor experiences an unsteady flow phenomena surge and rotating stall. These instabilities can cause noise nuisance and critical operating conditions with strong dynamical loading on the blades. Therefore, they cannot be tolerated during compressor operation. With the reduction of the compressor inlet flow, a rotating stall occurs in the impeller, or diffuser or scroll. If the compressor inlet flow continues to reduce, the rotating stall eventually will drive the compressor into a surge [27–32]. A rotating stall is an unsteady and three-dimensional flow phenomenon. CFD simulation is still a big challenge. The flow range of a centrifugal compressor can be extended by allowing gas to bleed from a ring of holes or a circular groove port around the compressor casing at a point slightly downstream of the compressor inlet. This type of the compressor called ported shroud compressor. Ported shroud forces a simulation of impeller stall to occur continuously, allowing some air to escape at port location inhibits the onset of surge and widens the operating range. The flow inside port is unsteady and complicated when compressor stalls. CFD guides ported shroud design is still very time consuming and less reliable. Some design practices for a ported shroud impeller casing [33] are discussed for improving the compressor operational range.

INDUSTRIAL CENTRIFUGAL COMPRESSOR

Centrifugal compressors are widely used in automotive, marine turbocharging, oil and gas, aerospace, and distributed power applications because of their compact design and high stage pressure ratio. With different types of applications, the structural characteristics of the compressors have two basic types, that is, horizontal split and vertical split, as shown in Figure 1. Horizontal split type compressors are applied for low-to-medium pressure service, as shown in Figure 1(a). This type of casing is split along the rotor shaft and bolted at the split line. The bearing and seal sections allow easy disassembly and assembly via the inspection cover, without having to remove the upper casing. A vertical split compressor is easy to access the gears, bearings, seals, and be repaired on site. However, due to the large crossing area in the splitting surface, it is difficult to prevent gas and lubricant oil leakage. A vertical split compressor is applied for medium- and high-pressure service, as shown in Figure 1(b). This type of compressor consists of an inner casing and an outer casing. The inner casing forms a single unit with the head, bearing, and seal and is fixed to the outer casing by shearing. The nozzle can be attached to the top, bottom, or side in accordance with client specifications. Both bearings and seals can be inspected without removing the inner casing. However, the manufacturing cost and installation cost may be higher than for a vertical split compressor. It is also not easy to access the gear, bearings, and seal. For combining the advantages of both split types, a hybrid split has become popular. For some applications, the gearbox can be a vertical split, and the compressor stage can use a horizontal split, as shown in Figure1(c).

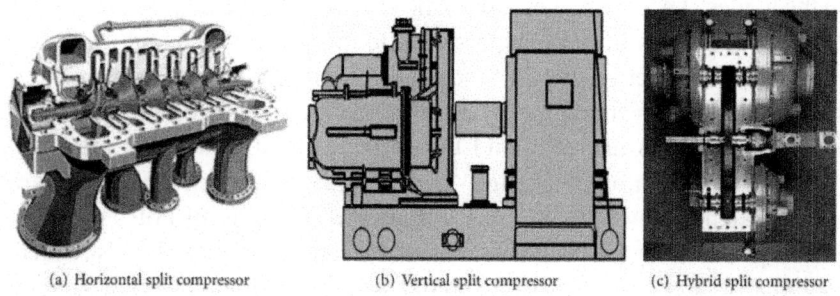

(a) Horizontal split compressor   (b) Vertical split compressor   (c) Hybrid split compressor

Figure 1: Compressor split types.

IMPELLER DESIGN METHODOLOGIES

The impeller is a key component to influence overall performance of a centrifugal compressor [5]. The efficiency of centrifugal compressors has increased dramatically, especially low-pressure ratio centrifugal compressors. A major challenge for a centrifugal compressor design is to keep a high efficiency level at a state of the art and to increase the compressor operating range [27–33]. Increasing the compressor operating range without sacrificing compressor peak efficiency is difficult to achieve. Aerodynamic engineers not only need to understand the surge physics but also need to apply design experience to the design. Another important objective for impeller design is to reduce the manufacturing cost. Manufacturing cost could be reduced when designs for manufacturability are effectively considered. The impeller should meet requirements to be easily withdrawn from a casting mold without destruction and disassembly of the mold for a casting impeller. This requires the lean angle of the blade to change linearly with the impeller radius and axial direction, as shown in Figure 2. All these considerations for design will help the final design to meet the compressor design target with fewer design iterations.

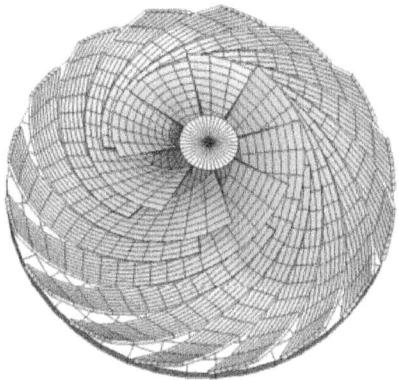

Figure 2: Impeller for manufacturing.

Different designers may have different methodologies for impeller developments. What kind of area distributions, curvatures, velocity, or pressure profile will lead to a good design is strongly dependent on the designers' practice and experience. Two totally different design philosophies could produce similar performance. For example, two impellers designed by Garret and Pratt Whitney [7], as shown in Figure 3, had different shapes with similar performance at the design point. The impeller designed by the author also presented different features, which also provide a good performance. It is shown that, if design follows basic design guideline, a wide range of solutions to the design can be used.

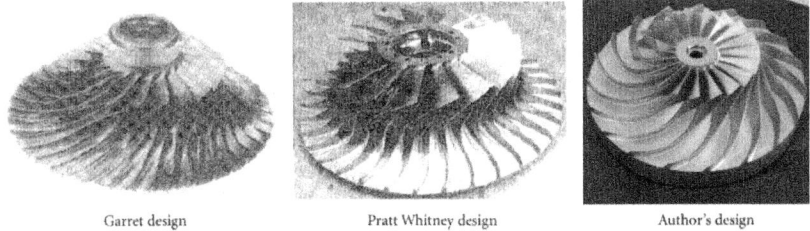

Garret design   Pratt Whitney design   Author's design

Figure 3: Different impeller designs.

The authors' design, as shown in Figure 3, is an example of a recently developed single-stage centrifugal compressor. At the design point, the total to static stage pressure ratio was about 3.7, and the flow coefficient was about 0.12. The running clearance at the impeller tip was 4.5% of the impeller exit blade width. Six builds were assembled and tested based on the ASME PTC-10 test procedure [34]. The compressor performance obtained from an average of six build tests is shown in Figure 4. The differences of test results for different builds for adiabatic efficiency and head coefficient were within ±0.5% and ±0.75%, respectively. The test uncertainties for total pressure (in psi), static pressure (in psi), and temperature (in Fahrenheit) were ±0.25%, ±0.2% and ±0.5%, respectively, based on uncertainty analysis [14]. Test results showed that the compressor performance was encouraging at both the design and the off-design point. The design met the low-cost target and allows large manufacturing tolerances. The insensitivity of the impeller surface finish and large tip clearance makes it easy to assemble.

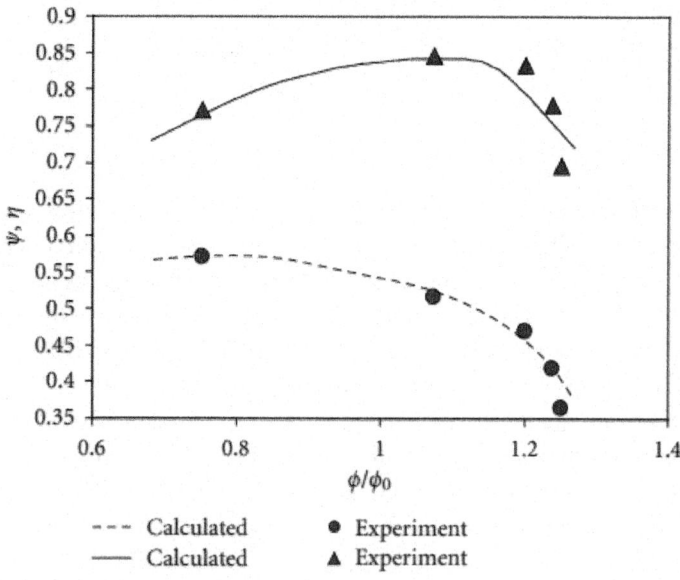

Figure 4: Single stage compressor efficiency and head coefficient versus flow coefficient (upper curve is the adiabatic efficiency, and lower curve is head coefficiency).

IMPELLER GEOMETRY

The initial design of a centrifugal compressor always begins with requirements from customers or marketing analysis. Designers select basic configurations and provide basic performance to customers or marketing by using their experience data. Aerodynamic designers also need to provide an estimation for the compressor's basic geometry to engineers in other disciplines. For example, rotor-dynamic engineers and bearing designers rely on the impeller geometry information to perform their work. Although the basic geometry design is not intended to yield optimization of the impeller, it can accelerate the overall design process and reduce the development cost.

Before aerodynamic designers determine the basic impeller geometry, the rotational speed of the impeller needs to be selected. If there are no special requirements for rotational speed, we normally optimize rotational speed based on the Balje's charts [35, 36] by using optimal specific speed. Although Balje's charts are not very accurate tools, they are sufficient enough to provide the initial estimate for impeller geometries.

During the initial design, the important information needed for bearing designers and rotor-dynamic engineers is impeller weight. Aerodynamic designers can estimate impeller sizes based on the required gas flow, pressure ratio, and impeller rotational speed. Our design practices showed that the weight of the impeller is the function of the impeller diameter. Figure 5 summarizes the relationship between the impeller diameter and the weight for sixteen ASTM A564 stainless-steel unshrouded impellers. The impeller weight mainly is determined by the impeller disk; the blades only contribute a very small portion of the weight. Therefore, we plotted impeller weight and diameter relation in one figure for all designed unshrouded impellers with different blade counts and with or without splitters.

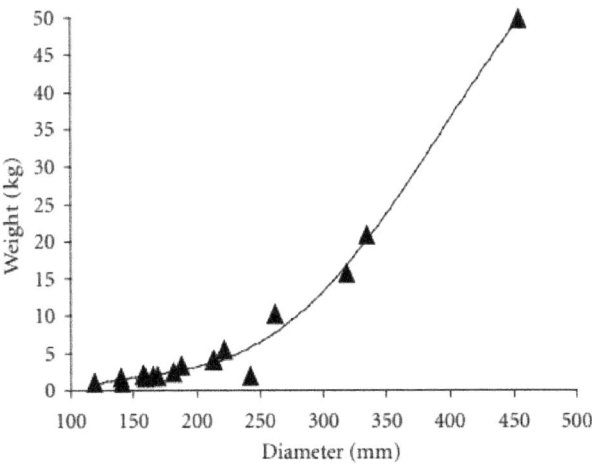

Figure 5: Impeller weight with diameter.

In an initial stage of compressor design, selections of the impeller inlet, the outlet velocity vectors, and the choice of blade numbers are the key initial design decisions. Velocity vectors may be obtained through a mean-line program. The experience data show that both the inlet blade numbers and the exit blade numbers are a function of stage pressure ratio. Relationships between the numbers of blades and the stage pressure ratio are shown in Figures 6 and 7 for with and without splitter impellers, respectively.

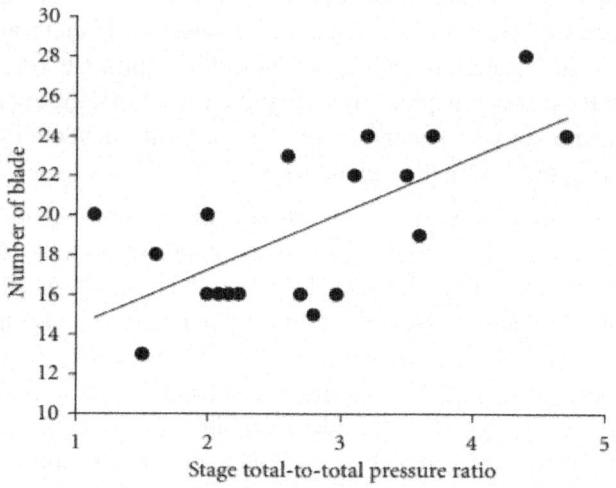

Figure 6: The number of blades at the inlet versus stage pressure ratio.

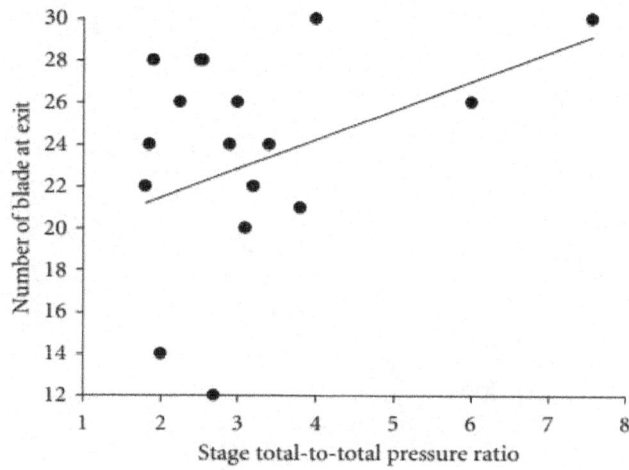

Figure 7: The number of blades at the exit versus stage pressure ratio for split impellers.

In general, high-stage pressure causes a blade-loading increase, and the impeller needs more blades to distribute loading. Variations of the numbers of blades at a similar pressure ratio were due to the size of the impellers. For a smaller-sized impeller, manufacturing capabilities may limit the number of blades. Impeller sizes plotted in Figures 6 and 7are in the range from 2 inches to 45 inches. The machine performance requirements and manufacturing feasibilities are factors to determine whether to use splitters or not.

The inlet blade height is determined by the design inlet flow rate and the impeller hub radius. The inlet hub radius is determined by the attachment of the impeller. For overhung impellers, the inlet hub radius normally is selected in a range between 10% and 20% of the impeller tip radius. For the shaft and bolt through the impeller, selections of the inlet hub radius are based on stress requirements.

The blade thickness at the inlet and the discharge was determined mainly by tensile and bending root stresses at the leading edge and the blade exit. FEA calculations and stress tests showed that blade root stresses are mainly caused by the centrifugal force. The blade high was a key factor to impact the blade root stresses. The mean-line thickness at the inlet and the exit was determined by the blade heights at the inlet and the exit, as shown in Figures 8 and 9. Experience showed that the blade thickness changed linearly with the blade height.

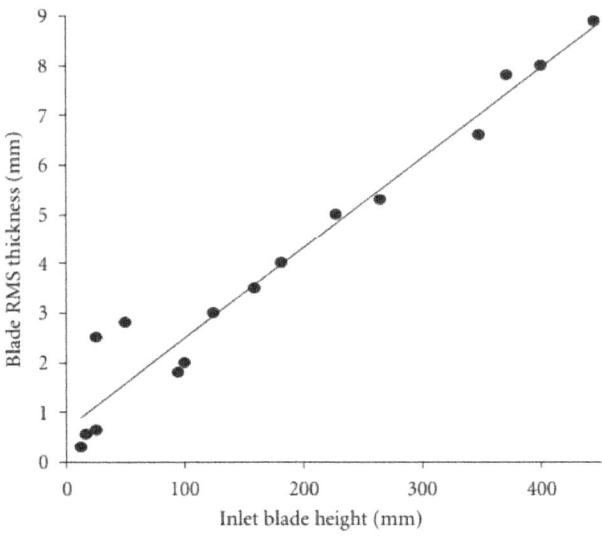

Figure 8: Relationship between impeller inlet RMS thickness and inlet width.

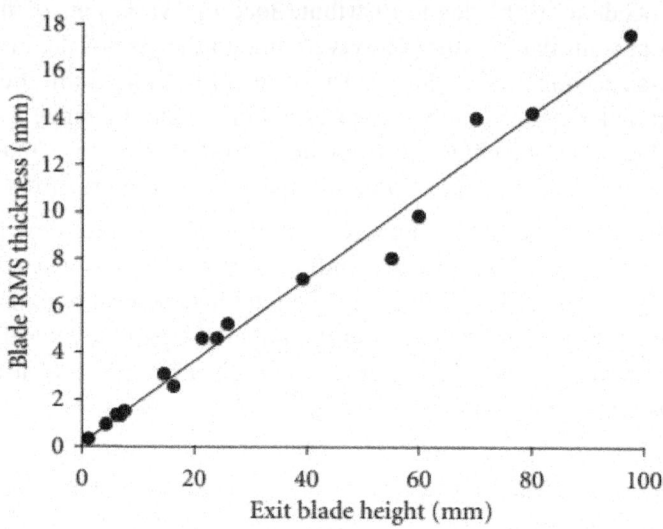

Figure 9: Relationship of impeller exit RMS thickness and tip width.

The three-dimensional features of the impeller blade are dependent on the engineers' experience and on stress limitations. The modern impeller is normally a three-dimensional design. The wrap angle, lean angle, and back sweep angle use a larger value than those in the past. The large wrap angle can reduce the camber of the blade but increase the frictions of the fluid. A large lean angle permits blade design at all blade sections with the desired shape. Leaning the blades creates a back-sweep and retains purely radial fibers, which are beneficial for bending moments. Experience showed that impellers with a back sweep generally have high efficiency.

REYNOLDS NUMBER AND SURFACE FINISH

The Reynolds number or the Ross by number has significant impacts on the impeller maximum surplus value. Fundamental fluid dynamics theory [37, 38] shows that the flow inside a pipe for a different Reynolds number represents different flow patterns. This is also true for flow inside impeller blades. Experience showed that if there is a flat velocity profile at the inlet between two impeller blades or diffuser vanes, the flow development along the flow channel presents different profiles with different Reynolds numbers. For low Reynolds number flows, the exit velocity profile is almost parabolic and only with a small portion of a flat profile. For high Reynolds number flows, the exit velocity profiles have large flat profiles.

The peak meridional velocities for high Reynolds number flows are normally located at the hub pressure sides of the blade due to potential flow effects. Low Reynolds number flow regions are located near the suction side of the blades. The viscous jet and wake interaction causes flow separations. The Reynolds number also strongly influences secondary flow patterns. Increasing the Reynolds number increases the strength of the clockwise secondary passage flow circulation. Reducing pressure on the suction velocity gradient increases the flow circulation of the counter clockwise secondary flow. The optimum design should try to offset each other to minimize the secondary flow losses.

The machining and casting of centrifugal compressor impellers and other components result in an inherent surface roughness. The sizes and forms of roughness depend on the manufacturing process. The levels of the surface finish represent the manufacturing cost. It is very important to balance manufacturing cost and performance. Surface finish requirements for different designs have different requirements. A detailed discussion on surface finish and Reynolds number can be found in reference [39, 40]. Loss due to the surface finish can be represented as wall friction. Wall friction is the function of Reynolds number and can be written as

$$\frac{1}{\sqrt{f}} = 1.74 - 2\log_{10}\left(\frac{k}{B_2} + \frac{18.7}{\mathrm{Re}\,\sqrt{f}} \right).$$

(1)

This equation can be solved by using a simple computer program or spreadsheet.

IMPELLER AERODYNAMIC DESIGN

One important guideline for impeller aerodynamic design is to set a reasonable diffusion ratio. The diffusion of the impeller can be represented by velocity ratio, diffusion factor, and relative Mach number ratio. The ratio of the relative Mach number was used in this discussion because it can avoid the one-dimensional assumption at the inlet. MR2 is defined as the ratio of the relative Mach number at the impeller inlet to the average Mach number at the impeller exit. Figure 10 shows the upper and lower boundaries for maximum deceleration likely to be achieved for two-dimensional and three-dimensional impellers [8]. The experience data fell inside the theory boundaries, and the Mach number ratio MR2 fell between 1.15 and 1.4, giving a good overall performance. The upper boundary of MR2 of 1.4 for an industrial compressor and 1.7 for jet engine impellers are reasonable expectations. Our experience also indicated that a large diffusion might cause a large loss. The ratio of the

Mach number can be selected within a large range. An important factor to impact the selection of diffusion level is the inlet Mach number. Figure 11 is a relationship between the relative velocity ratio, the incidence, and the inlet relative Mach number for a typical industrial impeller. It can be seen that diffusion is not an absolute parameter, which influences the stall of the compressor. It is worthwhile to point out that this test impeller was stalled first at the inducer. The inducer shroud velocity represented the rotational speed.

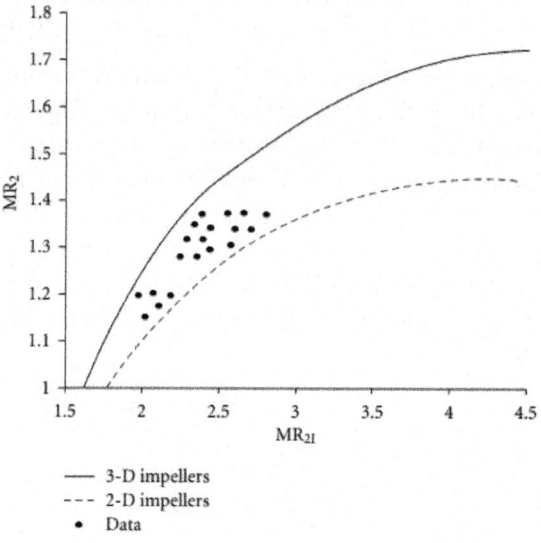

Figure 10: Relationship of MR_2 versus MR_{2I}.

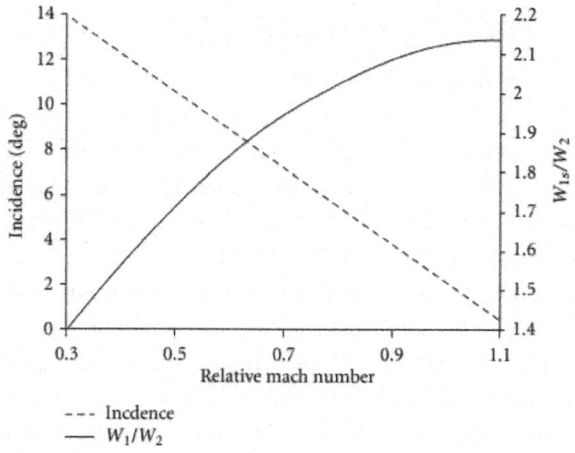

Figure 11: Stall incidence and relative velocity ratio versus Mach number.

Traditionally, the impeller inlet incidence is set to zero at the design condition. Modern impeller designs need not only to consider maximum efficiency at the design point, but also to consider the manufacturing cost and the off-design performance for the whole operating range [8]. Inlet blade angles are not necessary the same as the inlet relative flow angles. Experience data in Figure 11 shows that changes of the inlet flow incidence impact both the efficiency and the operating range of the impeller. Figure 12 shows that little negative incidences could raise the impeller operating range. However, when the negative incidence increased to a certain level, the operating range did not enlarge and efficiency dropped significantly. The impeller design should avoid this situation.

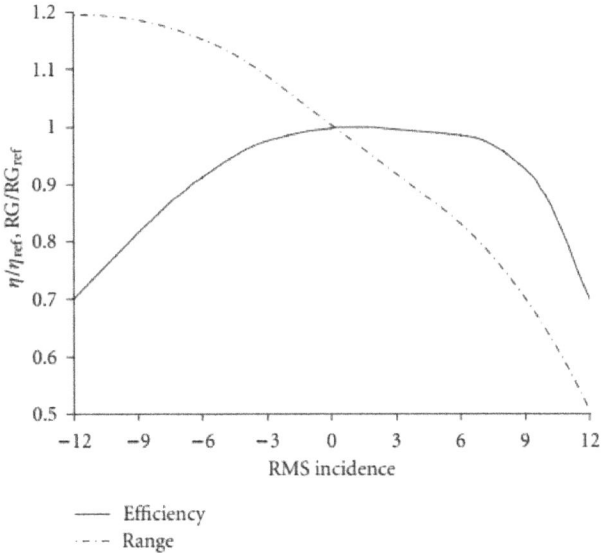

Figure 12: Incidence versus efficiency and range.

Estimations of the impeller exit width are critical for both the primary performance estimation and the basic dimension setup. The major impacts of the impeller blade exit width are the flow capacity and the pressure ratio of the stage. It is difficult to calculate the impeller exit width accurately in a simple way. The Rodgers diffusion factor equation [41] provided a good estimated value for the impeller exit width (B_2). If the mean meridional blade length can be estimated asthen the impeller exit width can be estimated as

$$L = \frac{2\pi \left(r_2 - r_1\right)}{4},$$

(2)

then the impeller exit width can be estimated as

$$B_2 = \frac{10\,(D_2 - D_{s1})}{1 + (W2/W_1)} \left[DF - 1 + \frac{W2}{W_1} - \frac{\pi D_2 C_{r2}}{2LzW_1} \right]$$
$$- \frac{(D_{s1} - D_{h1})}{2}.$$

$$(3)$$

Secondary flows inside the impeller are caused by an imbalance of the static pressure and the kinetic energy. One of the typical secondary flows, the horseshoe vortex, has been well documented. It is shown that the strength of the secondary flows is governed by the vortex starting conditions. The further development of the vortex is determined by the conservation of angular momentum. The impeller meridional blade profiles influence the secondary flow loss level and the laminar viscous dissipation function can estimate the secondary flow loss due to the blade profiles [8]. We have

$$\Delta H_l = N \int \mu \left[2\left(\frac{\partial u}{\partial x}\right)^2 + 2\left(\frac{\partial v}{\partial y}\right)^2 + \left(\frac{\partial u}{\partial y} + \frac{\partial v}{\partial x}\right)^2 \right] dV.$$

$$(4)$$

Tip clearance cannot be avoided for unshrouded impellers. Bearing clearances and manufacturing tolerances of the impeller and the intake ring control the minimum impeller tip clearance. The minimum tip clearance is normally defined at the maximum rotational speed with hot weather conditions for most motor-driven compressors. For compressors installed in the same shaft with a gas turbine, the minimum compressor tip clearance was estimated when the compressor was operating at maximum rotational speed with hot weather and with the machine overall net axial thrust load towards the compressor. The tip clearance increases quadratically with the impeller rotational speed if other operating conditions do not change. The tip clearance impacts the overall compressor performance because it increases the magnitude of the secondary flow inside the impeller blades and produces strong tip vortices. The tip clearance flow transports a low momentum fluid from the suction side to the pressure side of blades. The circumferential center of the secondary flow is dependent on the size of the tip clearance. Secondary vortices are located near the shroud side for small clearance, whereas secondary vortices may spread to the center and even the hub of the flow channel for a large tip clearance impeller. The clearance distribution affects the wake formation and the location at the impeller exit. A large clearance at the leading edge results in a low-energy center close to the suction side of the blade. Reducing the clearance at the leading edge, the wake moves towards the pressure side of the blade. The tip clearance setting depends on the compressor's maximum surplus value. There are several methods to reduce the tip clearance losses. Figure 13 shows that variable clearance could significantly improve the stage efficiency of compressors. The tip clearance changes the compressor stage head and capacity. Test data indicated that the head coefficient almost changed linearly with the tip clearance, as shown in Figure 14, and the flow coefficient followed a secondary order curve with clearance, as shown in Figure 15.

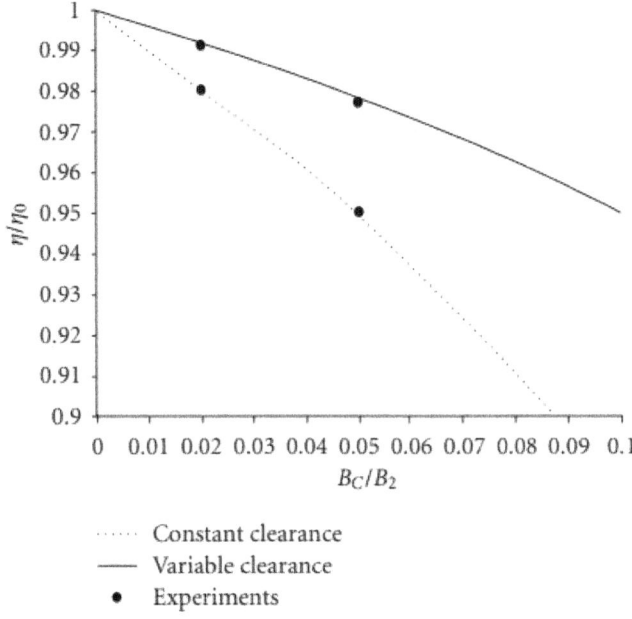

Figure 13: Compressor efficiency change versus clearance.

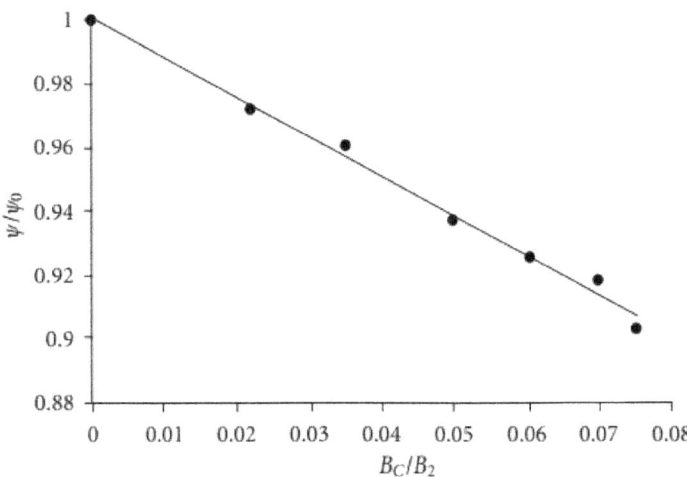

Figure 14: Head coefficient change with tip clearance.

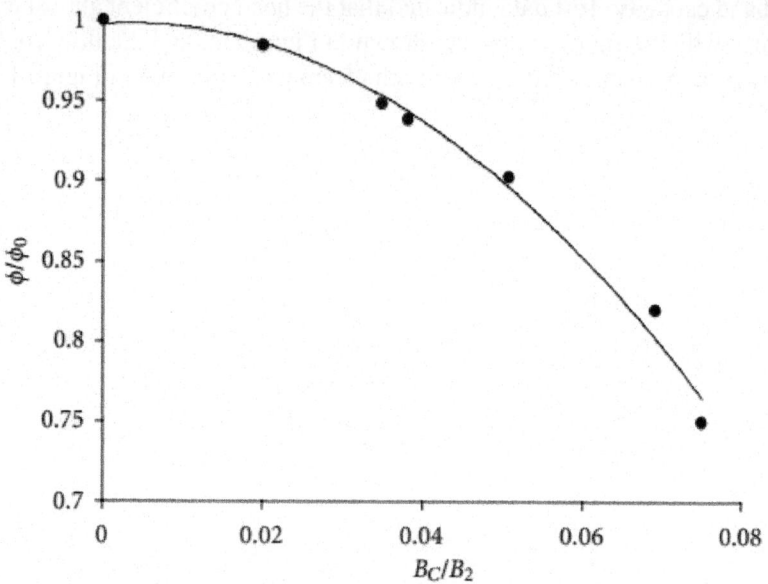

Figure 15: Flow coefficient change with tip clearance.

With the manufacturing technology improvement and design system improvement [42, 43], modern impellers always are designed in a three-dimensional shape. It is very important to understand the impact of the three-dimensional feature on performance and structure. Lean is one of the critical three-dimensional features. In general, both negative and positive lean improves the peak efficiency. One of the design examples for lean effects is shown in Figure 16. Negative lean has the best peak efficiency, while positive lean has a wide operating range. It can be seen from Von Mises stress contour plots in Figures 17 and 18 that the Von Mises stress is the highest for a negative four-degree lean on a suction surface, and the stress is the lowest for a four-degree lean. However, the pressure side stress contours show the lowest leading edge stress area for a negative four-degree lean design compared with other cases. The negative lean has maximum stress for both the leading edge and the trailing edge while the positive lean has the lowest maximum stress in all locations. The highest stress areas for the positive lean are less than the negative lean and no lean cases. The bore stress and back face stresses are similar because the blade thickness distribution is similar. It can be seen that a small positive lean can reduce the peak stress.

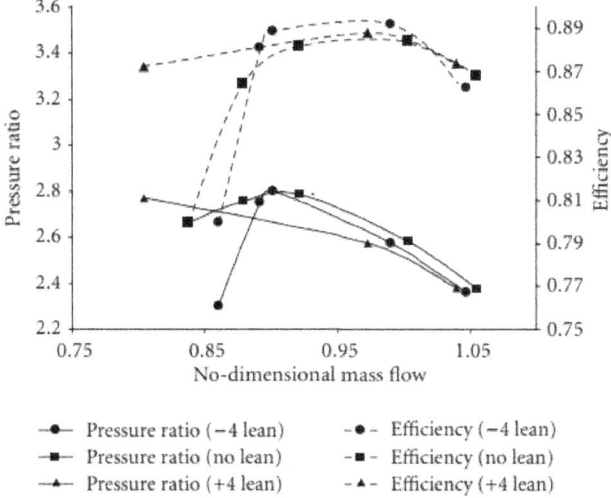

Figure 16: Compressor performance for a lean blade.

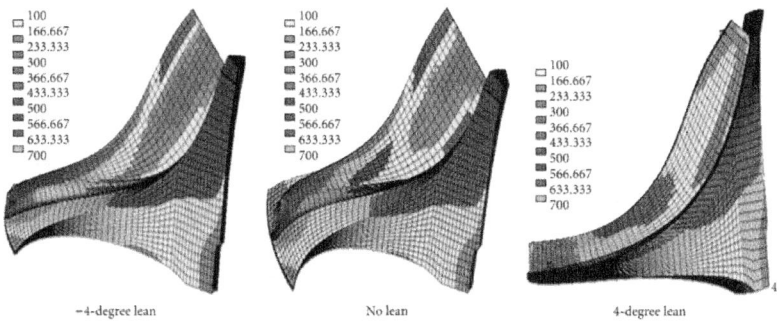

Figure 17: Von Mises stress at the suction side of the wheel.

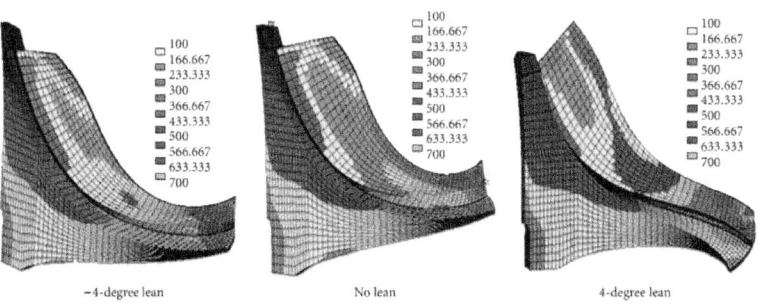

Figure 18: Von Mises stress at the pressure side of the wheel.

IMPELLER FLOW CUT

For large volume low-cost compressors, for example, automobile turbochargers, the impeller flow cut is always used for meeting different flow and pressure ratio requirements with minimum manufacturing cost. A flow cut is defined as a change to the impeller blade height or a change to the shroud contour while following the same hub contour and blade angle definition as the original impeller design. This allows a single manufacturing method to be used for the base impeller, plus a machining operation to adjust the blade height and the flow capacity of the impeller. Achieving the objectives for high efficiency and end users' cost of a compressor installation are always in conflict with each other. The flow cut of a compressor stage is one of the important activities of centrifugal compressor manufacturers to achieve a certain performance level with minimum cost. A theoretical equation provides the basic information, but the empirical performance effects on the compressor stage are very important to the compressor manufacturers. Figure 19 shows the original impeller (Figure 19(a)), the shroud contour cut (Figure 19(b)), and a diameter cut (Figure 19(c)). Modern impeller design normally has a bigger back sweep angle, as shown in Figures 2 and 3. The diameter cut, as shown in Figure 19(c), affects the impeller performance at a significant level. Therefore, the diameter cut is less popular than the shroud contour cut.

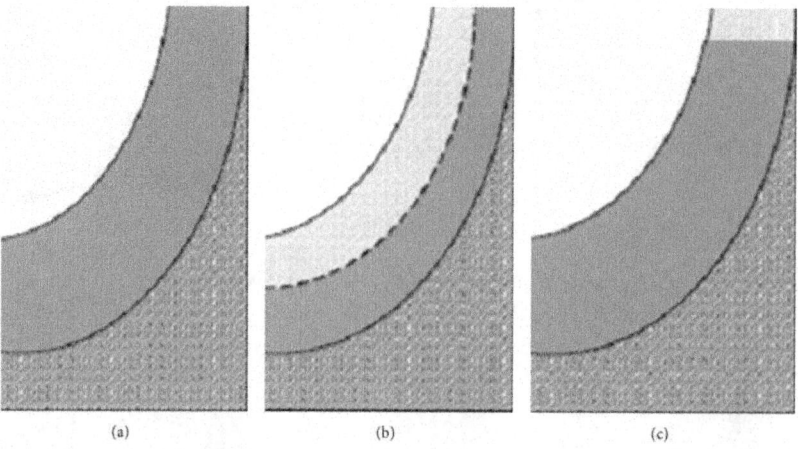

(a)   (b)   (c)

Figure 19: Flow cut of the impeller.

Here we present some test experience for both the shroud and the diameter contour cut. The shroud contour cut (or extend) is defined as the percentage of the local blade height. Most of time, manufacturers extend or trim the shroud contour the same percentage of the impeller inlet and exit blade heights. For large industrial compressors (choke flow larger than 1000 ICFM (0.472 m³/s)), the flow changes normally near linear to the contour change. The major impacts of the flow cuts on the compressor performance are the compressor stage efficiency and the surge flow or the operating range (RG). Figure 20 is the head coefficient at surge point changes with the impeller contours, where the surge point defined here is the operation point at which the compressor stage adiabatic efficiency is 55%. It is shown that the head coefficient at the surge increased as the contour increased. This may be because when the contour increases, the cold tip clearance remains the same during the tests for all the contours. For a larger contour, the tip clearance percentage relative to the impeller exit blade height is smaller, and the surge margin improves. This may be for the same reason. When the contour increases a little, the stage efficiency increases until the design penalty away from the design point is larger than the clearance loss reduction, the compressor efficiency starts to drop, as shown in Figure 21. Test results also suggest at an increased contour, a broader range of high efficiency exists. Testing also indicates that the point of peak efficiency is not proportional to the change in the area of the impeller inducer and the exit geometry. It is interesting to see that the flow corresponding to the peak efficiency increases as the impeller contour increased, as shown in Figure 22. It is important to make the flow cut to ensure the peak efficiency at the operating flow.

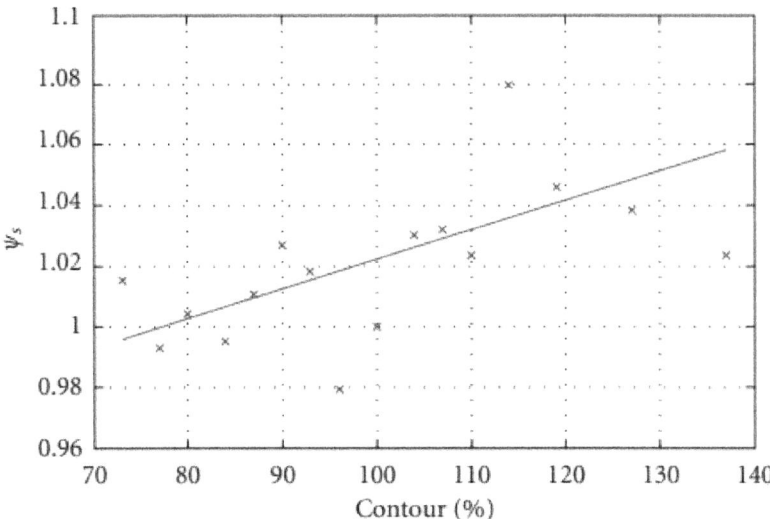

Figure 20: Head coefficient at the surge as a function of the contour.

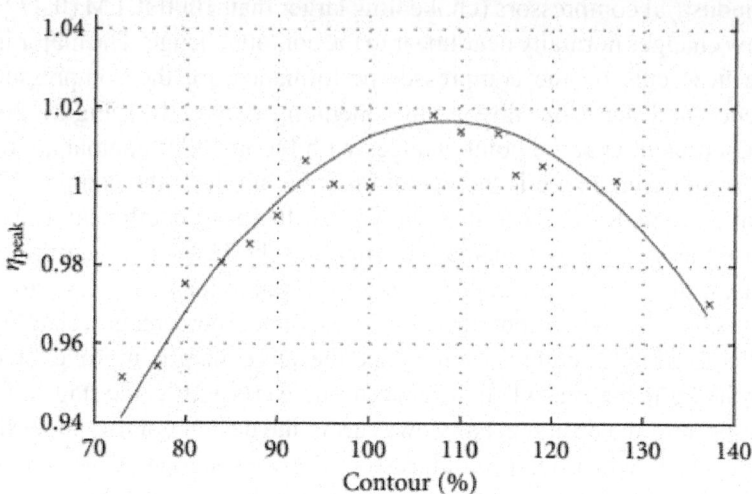

Figure 21: Peak efficiency change with flow cuts.

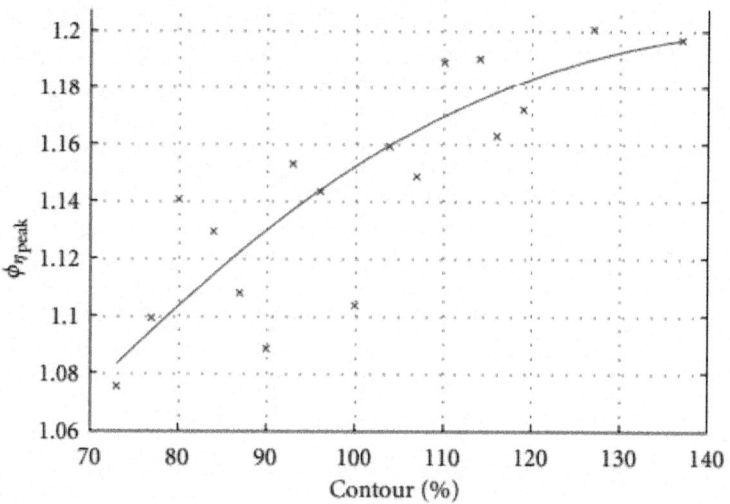

Figure 22: Flow coefficient at the peak efficiency versus the flow contours.

Although the diameter cut is less popular than the shroud contour cut, understanding fundamental information and basic compressor characteristics for the diameter cut is also very helpful. Figure 23 is the medium back-curvature impeller compressor (about 25 degree back-curvature) characteristics of a diameter cut. It can be seen that the diameter cut not only affect the peak efficiency but also affect the peak efficiency location. The impeller peak efficiency of a small diameter impeller is located in the lower pressure ratio operating point. It also can be seen that the compressor operating range is reduced after the diameter cut. It is interesting to notice that the surge boundary slope increases with the diameter decrease.

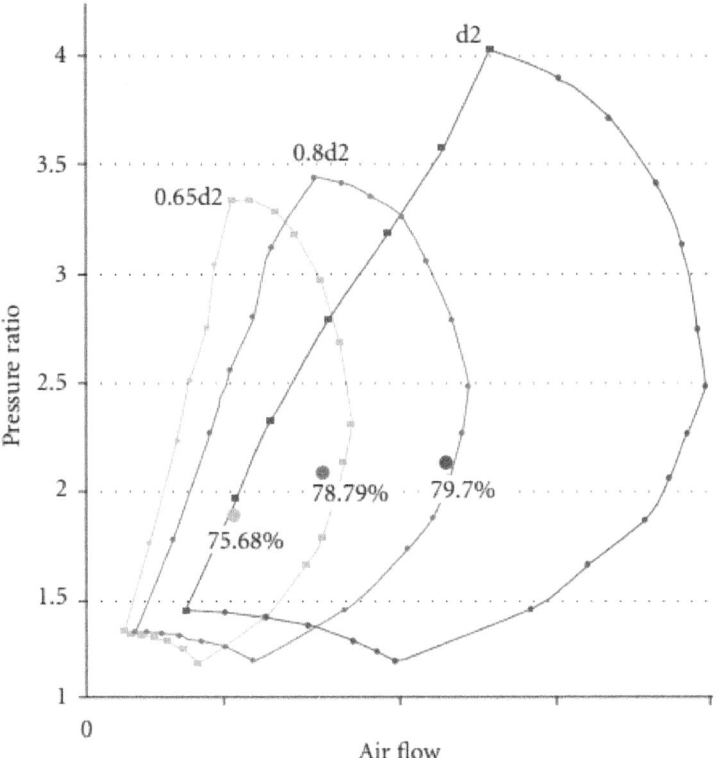

Figure 23: Performance characteristics of an impeller diameter cut.

PORTED SHROUD IMPELLER CASING

A compressor surge is a system-unstable phenomenon that is influenced by all components of the compressor. The physics of surge and stall are still not fully understood. We still cannot find any tool that can capture all features of the surge and stall, as shown in Figure 24. Many theoretical studies [27, 30, 31, 44] have focused on a better understanding of the surge and stall, but none can be used as a design tool yet. More theoretical work and experimental studies need to be done in order to incorporate stall in the design system. Designs for a wide operating range were mainly dependent on engineers' and manufacturers' experience and their understanding of the stall and surge. Rotating stall and surge are violent limit cycle-type oscillations in compressors, which result when perturbations (in flow velocity, pressure, etc.) become unstable. Originally treated separately, these two phenomena are now recognized to be coupled oscillation modes of the compression system—surge that is the zeroth order or planar oscillation mode, while a rotating stall is the limit cycle resulting from higher-order, rotating-wave disturbances, as shown in Figure 24. The compressor normally starts with a stall and then eventually become a surge, as shown in Figure 25. During the surge, the compressor can experience a reversing flow. System resistance releases after the flow reversal, and the flow starts flowing into the system. This surge cycle continues, as shown in Figure 25. The pressure variation rate with the mass flow rate variation is much larger than the near choke, as shown in Figure 26. To extend the surge margin, it is important to optimize the impeller, the vaned diffuser, the vaneless diffuser, and the volute design. The surge control can extend the surge margin further. The simplest the way is to have a bypass surge relief valve. When the centrifugal compressor pressure rises beyond a certain level, the valve bypasses the flow from the discharge to the inlet to prevent surge. A wise valve choice can help engineers pay for the rest of an advanced surge-control system without exceeding the budget. The valve needs to have a fast stroking speed when open, high capacity, low noise, and a very stable throttling control. For example, the oil and gas industry often chooses high-performance rotary valves.

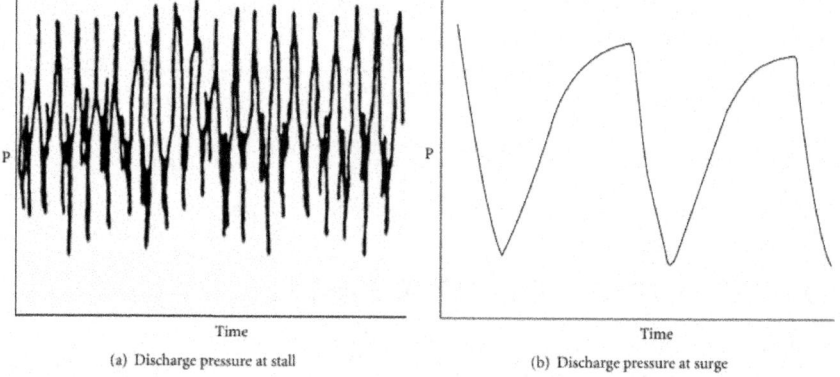

(a) Discharge pressure at stall

(b) Discharge pressure at surge

Figure 24: Stall and surge pressure variation with time.

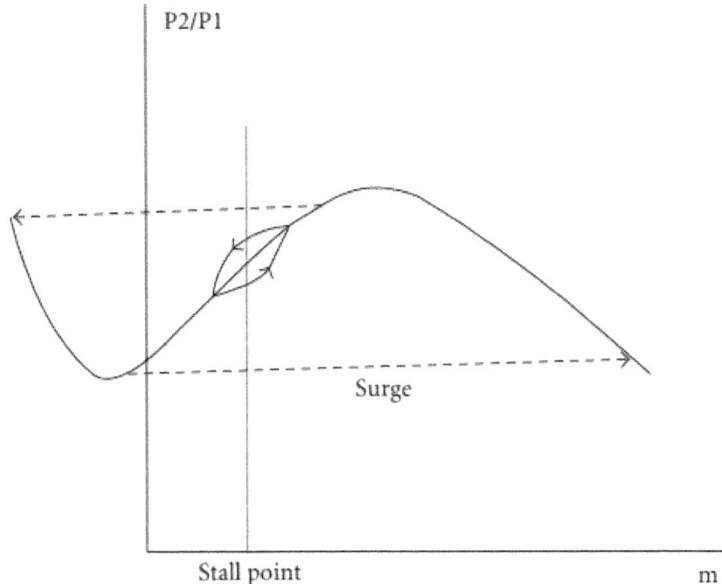

Figure 25: Surge of the centrifugal compressor.

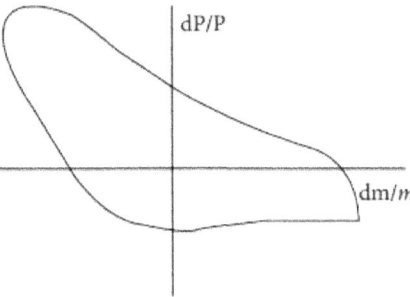

Figure 26: Variation of mass flow rate versus pressure fluctuations.

Casing treatments are other popular methods to extend the centrifugal compressor range. One of the key casing treatments is a ported shroud casing, as shown in Figure 27. The ported shroud design is entirely passive, having no moving parts, control valves, and so forth, The ported slot can be vertical to the inlet flow or angled, as shown in Figures 27(a) and 27(b). There are vanes or ribs to support the stationary shroud in the leading edge inducer region. The vanes or ribs are not only a support structure but also an aerodynamic device to the flow inside the port, bleeding out with minimum losses. The ported shroud passage provides the bleed path when the compressor is near the surge, as shown in Figure 28(a) and also a secondary air inlet to the impeller when the compressor needs more air, as shown in Figure 28(b). The vanes or ribs in the passage are tangentially slanted in the direction of the impeller rotation, to preferentially augment the airflow into the impeller at high speed, while in part-speed discouraging the airflow out of the impeller through the secondary inlet. The advantage of the ported shroud can increase in the part-load surge margin and increase the choke flow at full load, as shown in Figure 29. It can be seen that for a ported shroud compressor, at a lower mass flow the pressure ratio increases. This is because the separation and reverse flow were reduced near the inducer compared with an unported case. When the compressor is near choke, the impeller can bypass the impeller throat and draw an extra flow from the port, as shown in Figure 28(b). Figure 30 shows that the flow passed the port at different impeller mass flow conditions when the port width and location were optimized to have the best surge margin for a gas width equal to bg. It can be seen that the port recirculation flow increases when the compressor mass flow reduces. The port can pass as much as 40% of the impeller design flow. It can be seen from Figure 30 that the port recirculation flow reduces when the port width reduces. Figure 31 is the port width affecting the compressor performance for a ported compressor with the port location at 16% of the shroud meridional length from the impeller inducer. It can be seen that when the port location is fixed, the port width increases from 0 to $0.6b_2$, and both the surge margin and choke margin increase. The compressor adiabatic efficiency remains almost unchanged until the port width is larger than $0.1b_2$. The efficiency drops almost negligibly up to a port width larger than $0.2b_2$. Figure 31 also shows that the angle port (70 degree with axial direction) has an advantage compared with a straight port. The port design goal is to improve the surge and choke margin without greatly affecting the compressor performance. In some applications, for example, the automobile and aviation industry, the surge margin, and choke margins are both defined by certain efficiency levels. If the efficiency is too low, the system cannot perform properly. For those applications, the ported shroud applications are very popular. The ported shroud not only can extend the operational margin but also can improve the efficiency near the choke and surge. For those applications, the compressor map for a ported shroud demonstrated more advantages than an unported compressor, as shown in Figure 32.

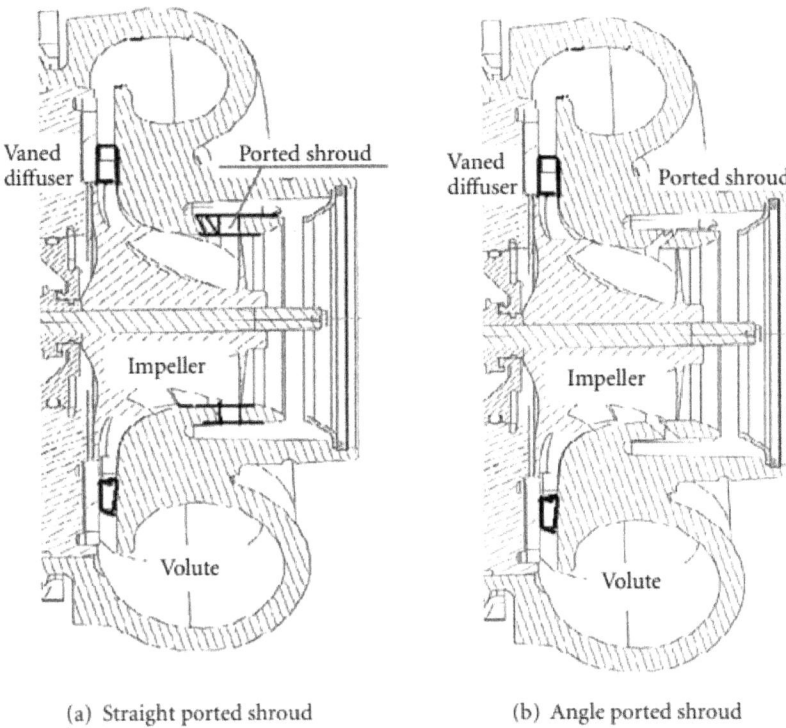

(a) Straight ported shroud   (b) Angle ported shroud

Figure 27: Ported shroud compressor.

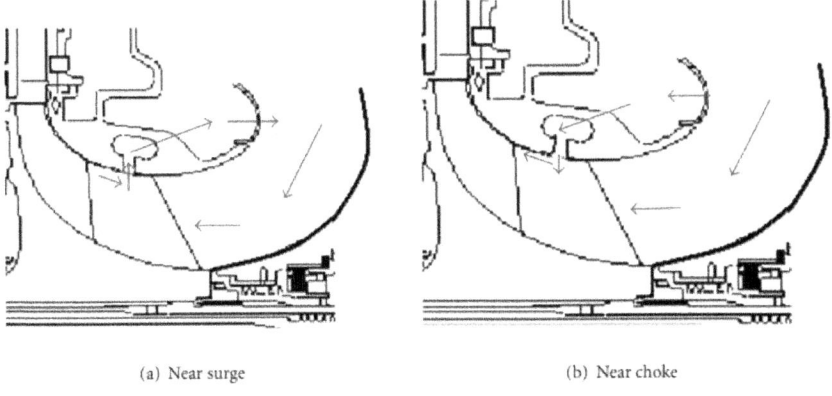

(a) Near surge   (b) Near choke

Figure 28: Ported shroud flows.

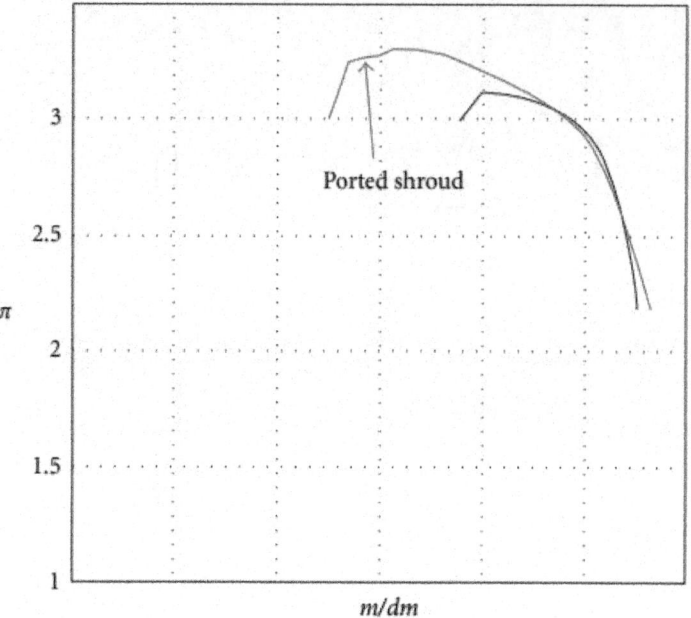

Figure 29: Compressor characteristic with and without ported shroud.

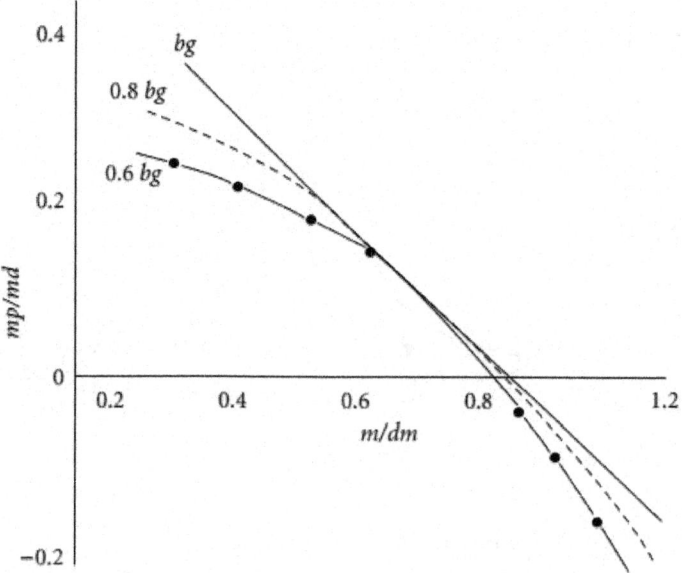

Figure 30: Flow inside bleed out from the port.

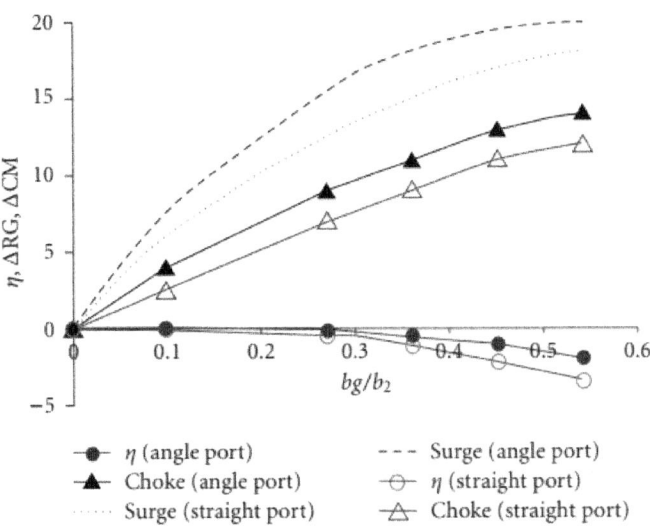

Figure 31: The compressor performance versus port width.

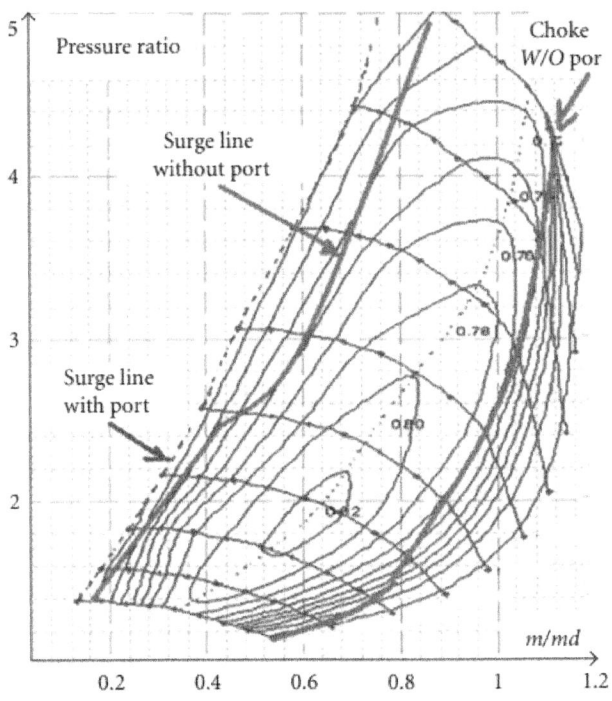

Figure 32: Ported shroud compressor map.

CONCLUSIONS

This paper provides several empirical consideration points in designing centrifugal compressors, focusing on the impeller design.(1)The different structure split compressors are discussed.(2)In an initial stage of compressor design, selections of the impeller inlet and the outlet velocity vectors and choice of blade numbers are the key initial design decisions.(3)Leaning the blades creates a back sweep and retains purely radial fibers, which are beneficial for bending moments. Experience showed that impellers with a back sweep generally have high efficiency. But lean blade design needs to consider the impacts of the impeller structure.(4)The velocity profile through the impeller blades largely depends on the flow Reynolds number. For low Reynolds number flows, the exit velocity profiles are almost parabolic and only with a small portion of the flat profile. For high Reynolds number flows, the exit velocity profiles have large flat profiles.(5)The tip clearance affects the overall compressor performance because it increases the magnitude of the secondary flow inside the impeller blades and produces strong tip vortices. (6)The tip clearance setting depends on the compressor's maximum surplus value. There are several methods to reduce the tip clearance losses. A variable tip clearance is one of the potential designs.(7)Compressor impeller flow cuts are widely used by compressor manufacturers. Some performance impacts need to be considered for doing the flow cut.(8)The ported shroud compressor has advantages for the compressor performance. Many new ported shroud structures have been proposed by many compressor manufactures [24]. But some fundamentals discussed in this paper can help designers to make the right decisions.

REFERENCES

1. M. P. Singh and J. J. Vargo, "Reliability evaluation of shrouded blading using the SAFE interference diagram," Journal of Engineering for Gas Turbines and Power, vol. 111, no. 4, pp. 601–609, 1989.

2. D. P. Kenny, The history and future of the centrifugal compressor in aviation gas turbine, SAE/SP-804/602, 1984.

3. P. D. Collopy, "Surplus value in propulsion system design optimization," AIAA-97-3159, 1997.

4. D. Japikse, "Decisive factors in advanced centrifugal compressor design and development," inProceedings of the International Mechanical Engineering Congress & Exposition (IMechE ‹00), November 2000.

5. C. Xu and R. S. Amano, "Development of a low flow coefficient single stage centrifugal compressor,"International Journal of Computational

Methods in Engineering Science and Mechanics, vol. 10, no. 4, pp. 282–289, 2009.

6. C. Rodgers and L. Sapiro, "Design considerations for high-pressure ratio centrifugal compressor," 72-GT-71, 1972.

7. R. H. Aungier, Centrifugal Compressors—A Strategy for Aerodynamic Design and Analysis, ASME Press, New York, NY, USA, 2002.

8. C. Xu, "Design experience and considerations for centrifugal compressor development," Proceedings of the Institution of Mechanical Engineers, Part G, vol. 221, no. 2, pp. 273–287, 2007. ·

9. C. Xu and R. S. Amano, "A hybrid numerical procedure for cascade flow analysis," Numerical Heat Transfer, Part B, vol. 37, no. 2, pp. 141–164, 2000.

10. C. Xu and R. S. Amano, "Flux-splitting finite volume method for turbine flow and heat transfer analysis," Computational Mechanics, vol. 27, no. 2, pp. 119–127, 2001.

11. C. Xu and R. S. Amano, "Computational analysis of pitch-width effects on the secondary flows of turbine blades," Computational Mechanics, vol. 34, no. 2, pp. 111–120, 2004.

12. C. Xu and R. S. Amano, "Numerical prediction of swept blade aerodynamic effects," in Proceedings of ASME Turbo & Expo, pp. 1–9, Vienna, Austria, June 2004, GT2004-53008.

13. C. Xu and R. S. Amano, "Computational analysis of swept compressor rotor blades," International Journal of Computational Methods in Engineering Science and Mechanics, vol. 9, no. 6, pp. 374–382, 2008.

14. C. Xu and R. S. Amano, "The development of a centrifugal compressor impeller," International Journal of Computational Methods in Engineering Science and Mechanics, vol. 10, no. 4, pp. 290–301, 2009. ·

15. C. Xu and R. S. Amano, "Computational analysis of scroll tongue shapes to compressor performance by using different turbulence models," International Journal of Computational Methods in Engineering Science and Mechanics, vol. 11, no. 2, pp. 85–99, 2010. ·

16. C. Xu and R. S. Amano, "Study of the flow in centrifugal compressor," The International Journal of Fluid Machinery and Systems, vol. 3, no. 3, pp. 260–270, 2010.

17. H. Krain, B. Hoffmann, and H. Pak, "Aerodynamics of a centrifugal compressor impeller with transonic inlet conditions," in Proceedings of the International Gas Turbine and Aeroengine Congress and Exposition, June 1995, 95-GT-79.

18. H. Krain, "Flow in centrifugal compressor," von Karman Institute for Fluid Dynamics, Ls 17, 1984-07, 1984.

19. D. Japikse, Centrifugal Compressor Design and Performance, Concepts Eti, Wilder, Vt, USA, 1996.

20. C. Xu and R. S. Amano, "On the development of turbomachine blade aerodynamic design system,"International Journal of Computational Methods in Engineering Science and Mechanics, vol. 10, no. 3, pp. 186–196, 2009.

21. E. Benini and A. Toffolo, "Centrifugal compressor of a 100 kW microturbine: part 2—numerical study of impeller-diffuser interaction," in Proceedings of the ASME Turbo Expo, pp. 699–705, June 2003, GT-2003-38153.

22. E. Benini and A. Toffolo, "Centrifugal compressor of a 100 kW microturbine: part 3—optimization of diffuser apparatus," in Proceedings of the ASME Turbo Expo, pp. 707–714, June 2003, GT-2003-38154.

23. J. Hoffren, T. Siikonen, and S. Laine, "Conservative multiblock Navier-Stokes solver for arbitrarily deforming geometries," Journal of Aircraft, vol. 32, no. 6, pp. 1342–1350, 1995. ·

24. T. Siikonen, "An application of Roe›s flux-difference splitting for k-ε turbulence model," International Journal for Numerical Methods in Fluids, vol. 21, no. 11, pp. 1017–1039, 1995. ·

25. M. M. Rai, "A relaxation approach to patched-grid calculations with the euler equations," Journal of Computational Physics, vol. 66, no. 1, pp. 99–131, 1986.

26. S. Niazi, A. Stein, and L. N. Sankar, "Development and application of a CFD solver to the simulation of centrifugal compressors," AIAA paper 98-0934, 1998.

27. W. Jansen, "Rotating stall in a radial vaneless diffuser," ASME Journal of Basic Engineering, vol. 86, pp. 750–758, 1964.

28. Y. Senoo and Y. Kinoshita, "Influence of inlet flow conditions and geometries of centrifugal vaneless diffusers on critical flow angle for reverse flow," ASME Journal of Fluids Engineering, vol. 99, no. 1, pp. 98–103, 1977.

29. H. S. Dou and S. Mizuki, "Analysis of the flow in vaneless diffusers with large width-to-radius ratios,"ASME Journal of Turbomachinery, vol. 120, no. 1, pp. 193–201, 1998.

30. S. Ljevar, H. C. de Lange, and A. A. van Steenhoven, "Two-dimensional rotating stall analysis in a wide vaneless diffuser," International Journal

of Rotating Machinery, vol. 2006, Article ID 56420, 11 pages, 2006.

31. K. Iwakiri, M. Furukawa, S. Ibaraki, and I. Tomita, "Unsteady and three-dimensional flow phenomena in a transonic centrifugal compressor impeller at rotating stall," in Proceedings of the ASME Turbo Expo, pp. 1611–1622, June 2009, GT2009-59516.

32. E. M. Greitzer, "Review—axial compressor stall phenomena," ASME Journal of Fluids Engineering, vol. 102, no. 2, pp. 134–151, 1980.

33. M. T. Barton, M. L. Mansour, J. S. Liu, and D. L. Palmer, "Numerical optimization of a vaned shroud design for increased operability margin in modern centrifugal compressors," ASME Journal of Turbomachinery, vol. 128, no. 4, pp. 627–631, 2006.

34. ASME, Performance Test Code on Compressors and Exhausters, PTC 10-97, 1997.

35. R. S. Amano and B. Sunden, Thermal Engineering in Power Systems, WIT Press, 2008.

36. O. E. Balje, Turbamachines, John Wiley & Sons, New York, NY, USA, 1981.

37. K. Brun and R. Kurz, "Analysis of secondary flows in centrifugal impellers," International Journal of Rotating Machinery, vol. 2005, no. 1, pp. 45–52, 2005.

38. R. L. Street, G. Z. Watters, and J. K. Vennard, Elementary Fluid Mechanics, John Wiley & Sons, New York, NY, USA, 1996.

39. T. Wright, "Comments on compressor efficiency scaling with Reynolds number and relative roughness," in Proceedings of the American Society of Mechanical Engineers, June 1989, 89-GT-31.

40. P. R. N. Childs and M. B. Noronha, "The impact of machining techniques on centrifugal compressor impeller performance," in Proceedings of the International Gas Turbine & Aeroengine Congress & Exposition, June 1997, 97-GT-456.

41. C. Rodgers, A diffusion factor correlation for centrifugal impeller stalling, 78-GT-61, 1978.

42. C. Xu and R. S. Amano, "A turbo machinery blade design and optimization procedure," GT-2002-30541, 2002.

43. D. P. Kenny, The history and future of the centrifugal compressor in aviation gas turbine, SAE/SP-804/602, 1984.

44. R. van den Braembussche, "Rotating stall in centrifugal compressors," VKI Preprint 1987-16, 1987.

Chapter 6

EXPERIMENTAL FITTING OF THE RE-SCALED BALJE MAPS FOR LOW-REYNOLDS RADIAL TURBOMACHINERY

Roberto Capata and Enrico Sciubba

Department of Mechanical and Aerospace Engineering, University of Roma "Sapienza", Roma 00184, Italy

ABSTRACT

With the increasing popularity enjoyed by ultra-micro scale turbomachinery, designers are often faced with severe challenges due to the substantial phenomenological difference between the low-Reynolds fluid-dynamics in rotating or strongly curved flows and the established knowledge acquired through decades of theoretical and experimental studies on medium and large-scale machines. The problem is complicated by the absence of an extended and reliable database that might be used for preliminary design and provide indications for scale-up or scale-down. As a result, custom-designed experimental campaigns are necessary that make the development of any new machine exceedingly costly. The situation has seen some improvement in recent years, after the publication of a sufficient number of experimental results and numerical simulations that pave the way towards a development of semi-empirical correlations. The purpose of this work is to present and discuss a preliminary and simple method to extend the currently available design maps into the small scale range (Re < 10^5) by introducing in the Balje charts an efficiency correction that depends on the specific speed n_s. The method results in a Stodola-like formula which originates a lower-than-standard Cordier curve on the classical Balje charts. A validation with some experimental results is also presented and discussed. Though the agreement is more than satisfactory, it must be stressed that the method provides only approximate results, and thus it must be considered as an evolving temporary solution, that needs to be updated as long as larger series of (numerical or physical) experimental results become available.

INTRODUCTION

In the early design stages, an (approximately) correct prediction of the performance of a turbomachine is critical, not only because it ensures that the type of components of the stage (impeller, diffuser and volute) are in line with current technology and with the design specifications, but also because it facilitates the subsequent final design task. The general performance maps available from manufacturers are usually limited to a small range of rotational speeds, pressure ratios and mass flow rates, and extrapolating on their basis is especially dangerous at the onset of a selection or design task. A comprehensive prediction procedure would enable the designer to analyze the performance of all the individual components already at the initial design stage, and would obviously result in an "optimal" design with a lower effort in terms of human and computational resources. Advanced computational fluid dynamics (CFD) procedures are better suited for later design stages, and one-dimensional prediction techniques continue to be the most practical method of predicting the performance of a stage or of one of its components [1]. Since the early works by Wiesner [2] and Stodola and Lowestein [3], one-dimensional prediction techniques and related semi-empirical loss correlations have been persistently developed [4,5,6,7]. The prevailing approach is to lump together losses of different type in the form of either a slip factor, a loss coefficient, or both. The downside of such a model is that the effects of individual losses are not clearly characterized and some phenomenological information is lost. This paper deals solely with polytropic losses in each component of a compressor (thermal phenomena linked to the non-adiabaticity of the flow being neglected) and lumps them together to calculate in a simple and convenient fashion their total extent. The model discussed here is based on the old idea of separating the losses in an "inviscid" and a "viscous" part, and subtracting them from the inviscid Euler work. In this perspective, the Reynolds number influence on the performance of turbomachinery has been the object of considerable attention in the technical literature, and several more or less general—and more or less reliable—models have been developed for both incompressible and compressible flow machines. Several proposals of more or less general models of an efficiency scaling of the type $\eta = f(Re)$ have been published, but all of the proposed models agree with the experimental data only within a limited range of configurations and fall therefore short of representing general design correlations. An acknowledged problem of the available formulations is the complexity and the non-crispness of the assumed phenomenological model. All approaches—except for the very detailed and costly 3-D, fully turbulent numerical simulations—lack sufficient accuracy and generality in separating Reynolds dependent and independent losses, and this makes it very difficult to quantify the Re-independent loss fraction, that clearly depends on manufacturing methods, tolerances and clearances. Most studies in this field are based on the classical but dated experiments carried out by Wiesner [2], Casey [8], Strub *et al.* [9] and others. The actual modelling varies somewhat from author to author, but a common starting point is the definition of a reference Reynolds number based on the rotor exit passage width:

$$Re_{b2} = \frac{U_2 b_2}{\nu}$$

(1)

Since the aim of this paper is to search for a possible correction factor to the Balje charts, our definition of the reference Re is at variance with Equation (1), and is based on the rotor outer diameter, like in Balje's approach [4]:

$$Re_{D2} = \frac{U_2 D_2}{\nu}$$

(2)

Some scaling formulations [2,6,8,9] include the influence of Reynolds number and surface roughness, but in this work only hydraulically smooth surfaces are considered, to eliminate roughness from the set of relevant variables: this choice was in fact almost unavoidable, given the scarcity of data available on the prototypal rotors in the ultra-micro scale range. We are very well aware of the influence of surface finishing in real machines, but suggest leaving the evaluation of its effects to specific lab-scale tests, easily performed on small-scale machines: we anticipate the possibility of reporting in a future work a possibly parametric correlation between efficiency and surface roughness. In the same line of reasoning, the effects of the non-adiabaticity [1] are neglected. In spite of these quite radical simplifications, the derivation of a "universal" formula is intrinsically difficult, due to the substantial difference in the flow phenomenology, so that each class of machines requires a specific analysis. The scaling formula suggested here has indeed (as most of the Re-correction formulae found in literature) the same structure as Wiesner's formula, the novelty residing in the fact that its coefficients have been derived via numerical simulations and validated by a series of laboratory tests. Such a double-calibration, so to say, makes it a valuable extension of the existing design guidelines. The formula is not intended to cover the entire range of "low-Re machines": in fact, its verified range of validity falls between $Re_{D2} = 1 \times 10^4$ and 1×10^5, which is common for ultra-micro machines ($Re \cong 10^5$). In the following, the suffix "2" in "D_2" is omitted.

REYNOLDS NUMBER EFFECTS

The general paradigm for all Re-correction formulae is the so-called Stodola [8] correlation:

$$\frac{1-\eta}{1-\eta_{ref}} = a + (1-a) \cdot \left[\frac{Re_{ref}}{Re} \right]^n$$

(3)

The coefficients a and n differ from author to author, as shown in Table 1, and represent the Re-independent loss fraction, which is are mainly related to curvature and rotation effects (relative vortex). Leakage losses, that are not really viscosity-independent, are usually accounted for by assuming that a is actually constant only in a small range of Re but that it may vary to a lesser extent with both geometry and manufacturing techniques even for machines with the same Re. It turns out, that a has in fact sometimes been found to be itself a function of Re, while there is a general agreement on the exponent n being inversely proportional to the Reynolds number. The usual practice, that we shall also follow, is to assume that the manufacturing process and clearances are in line with those specified by the Balje charts, so that a remains constant for the range denoted here as the "micro-machine scales", $i.e.$, $1 \times 10^4 < Re_D < 1 \times 10^5$. As for the exponent n, its value varies considerably from author to author, and though this in part may depend on the different intrinsic accuracy of the data [10], a more plausible explanation is that it, too, displays a Re-dependence: Wiesner [2] was the first author to postulate an explicit Re-dependence, and proposed a functional relation of the form:

$$n = c' \left(\frac{1}{Re/Re_{ref}} \right)^c$$

(4)

Remarkably, very few authors provide in their papers precise indications about the Re validity range (thence the "n.d." in Table 1): a careful inspection of the available graphs indicates that most formulae are applicable between 1×10^4 and 1×10^5, with Wiesner being the only author to supply a well-defined range: $5 \times 10^4 \div 5 \times 10^5$. As already reported by Balje, the pressure difference between suction and pressure sides causes a non-negligible amount of leakage over the blade tip.

Table 1: Summary of the most popular efficiency correction equations [2]

Year	Source	Inviscid loss fraction "a"	Viscosity-dependent loss fraction $(1 - a)$	Exponent n	Re range	Machine type
1925	Moody	0.25	0.75	0.33	n.d.	Propeller turbines
1930	Ackeret & Muhlemann	0.50	0.50	0.20	n.d.	Hydraulic turbines
1942	Moody	0.00	1.00	0.20	n.d.	Pumps
1947	Pfleiderer	0.00	1.00	0.10	n.d.	Pumps
1951	Davis, Kottas & Moody	0.00	1.00	Variable	n.d.	All turbomachines
1954	Hutton	0.30	0.70	0.20	n.d.	Kaplan turbines
1958	Rotzoll	0.00	1.00	Variable	n.d.	Pumps
1960	Wiesner	0.50	0.50	0.10	$5 \times 10^4 \div$ 5×10^5	Radial compressors
	Fauconnet	0.24	0.76	0.20		Radial compressors
1961	O'Neil & Wickli	0.00	1.00	Variable	n.d.	Radial compressors
1965	ASME Code PTC-10	0.00	1.00	0.20	n.d.	Axial compressors
		0.00	1.00	0.10		Radial compressors
1971	Mashimo *et al.*	0.25 min	0.75 max	0.20	n.d.	Radial compressors
1974	Mashimo *et al.*	0.15–0.57	0.43–0.85	0.20–0.50	n.d.	Radial compressors

This effect depends on the blade geometry, the clearance gap and the load. For what leakage is concerned, the non-Re related losses depend on other parameters, which scale according to different similarity rules, such as turbulence level, roughness effects, *etc.* Balje himself suggests the adoption of a "correction factor" for adjusting the turbo machinery efficiency. This consideration was included in our calculations.

CALCULATION OF THE CORRELATION COEFFICIENTS

It is quite apparent that Equations (3) and (4) are "complex" enough (*i.e.*, they subsume a sufficiently deep phenomenological model) to yield a reasonably accurate estimate of the losses for a particular class of machines such as the ultra-micro ones considered here. Due to the fact that ultra-micro turbomachinery has only been developed in relatively recent times, and in contrast to what happens for large-scale machines, a sufficiently large experimental database for coefficient fitting does not exist, which led us to the idea of adopting a "hybrid" fitting procedure, which consists in first deriving the coefficients via relatively inexpensive numerical simulations, and then verifying them by means of the available (limited) experimental database. To this purpose, a sizeable set of data was created by means of sufficiently accurate numerical simulations [11,12,13,14], to derive initial values for the constants a, c and c' in Equations (3) and (4). The simulations—discussed in detail elsewhere [11]—were performed on 3-D models in kinematic similarity using a commercial CFD simulation code, ANSYS/Fluent. The turbulence model was the k-ε realizable, with second order accuracy. Each model was meshed to ensure a $y^+_{max} \sim 5$, a necessary condition for adopting the enhanced wall treatment, since the quality of the grid has a relevant importance on the accuracy and stability of the numerical simulation. Commercial software allows the "plastering" of cell layers to the critical boundaries of the control volume, which are obviously, in this case, the wall surfaces of the hub, casing and blades. In these zones the usual practice is that of creating a completely structured boundary layer, specifying whenever possible both the height of the first row of cells and the "growth ratio", *i.e.* the rate that determines the height of the successive cells. In this process, the height of the first row of cells is usually determined via an empirical formula that gives the value of a wall-based local Reynolds number, denoted by y^+:

$$y^+ = u^* \cdot \frac{y}{v}$$

where $u^* = \sqrt{\frac{\tau_{wall}}{\rho}}$, τ_{wall} being the wall shear stress. For both turbine and compressor the procedure was the same: the stator control volume was split in several smaller sub-volumes, to achieve a more consistent set of faces and to better exploit the possibility of creating a locally more refined grid. The choice of the boundary conditions was made more difficult by the lack of conclusive experimental data: it was performed heuristically, starting from the preliminary sizing data, calibrating them by means of a first simulation, adjusting the values by iteratively resetting the outlet static pressure on the near-wake radial area downstream of the trailing edge. Through subsequent simulations the values of the inlet total pressure and temperature were refined as well in order to ensure conservation of the mass flow rate (the so-called "mass flow inlet condition" was adopted). The turbulent parameters were the hydraulic diameter D_{hyd} and the turbulence intensity $I = \sqrt{k}/v$. Rotational periodicity was imposed on all lateral channel surfaces. As discussed below, the resulting model retains its validity for compressors and turbines with diameter down to the order of millimetres, without the need for a re-calibration of the constants. The data extracted from the numerical simulations [11,12,14] of the ultra micro-compressor and turbine were used to calibrate the formula coefficients a, c, c'. The main dimensional characteristic of both component are shown in the simplified geometry of Figure 1 [11,13].

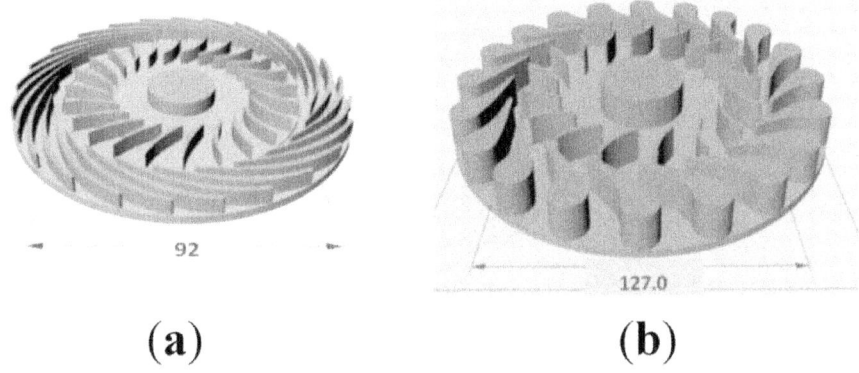

(a)   (b)

Figure 1: (a) 3-D sketch reference geometry for ultra micro gas turbine generator (UMGTG) compressor, where D_e = 92.0 mm [11,13]; (b) reference geometry for turbine, D_e = 127.0 mm [11].

To study the effect of the Reynolds number on the viscous-related losses, we simulated three models of the original device at different scales (1:3, 1:5 and 1:10). The data are reported in Table 2. The largest model is taken as the reference geometry (it falls within the range of the "commercial scales"). Table 2 indicates that the considered compressors and turbine are outside of the optimal n_s/d_s range. In fact, the Re_D of our 1:1 model falls within the range indicated by the Balje maps (Figure 2), where it displays a rather low efficiency, which was expected, because the adopted geometry is based on an "extruded" 2-D blade shape with zero twist.

Table 2: Data for different compressor and turbine models (all dimensions in mm)

Compressors							
Scale	D_2 (mm)	b_2 (mm)	ω (rad/s)	n_s	d_s	η_{st} (%)	Re/Re$_{ref}$
1:10	9.2	0.55	68745	0.38	5.7	52.5	1/10
1:3	27.6	1.65	65217	0.39	6.2	57.9	3/10
1:5	46.0	2.75	63376	0.40	7.2	59.2	2/10
1:1	92.0	5.50	57160	0.42	8.1	60.6	1
Turbine							
Scale	D_1 (mm)	b_1 (mm)	ω (rad/s)	n_s	d_s	η_{st} (%)	Re/Re$_{ref}$
1:10	12.7	1.31	68745	0.50	3.99	75.0	1/10
1:3	38.1	3.93	65217	0.47	4.20	76.1	3/10
1:5	63.5	6.55	63376	0.46	4.32	79.3	2/10
1:1	127	13.10	57160	0.41	4.91	80.0	1

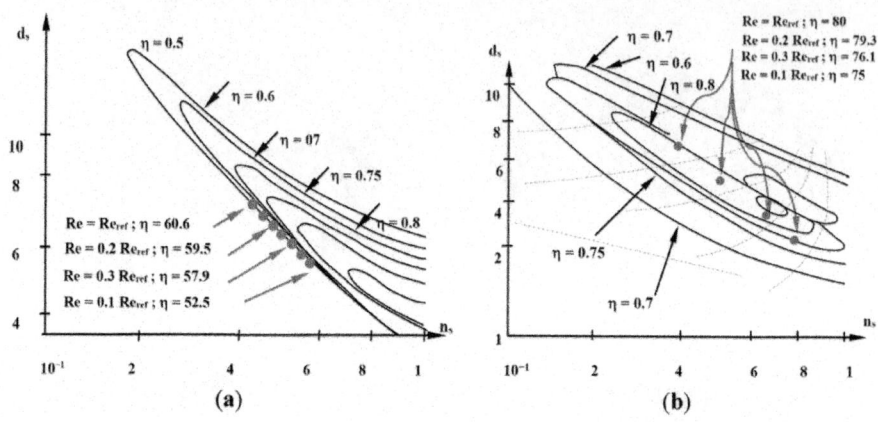

Figure 2: Computed efficiencies n_s for (a) turbine and (b) compressor . $Re_{ref} = 10^6$.

By fitting the computed data we obtained a value 0.50 for a, and respectively 0.25 and 0.084 for c and c'. More precisely, the value of a was calculated (to the third significant digit) using the method of the bisections for every single pair of data points. The input for the method are the simulation values, in an interval $[p,q]$, and the function values $f(p)$ and $f(q)$. The function values are of opposite sign (there is at least one zero within the interval). Each iteration performs these steps:

(1)   Calculate x, the midpoint of the interval, $x = 0.5 \cdot (p + q)$;

(2)   Calculate the function value at the midpoint, $f(x)$;

(3)   If convergence is satisfactory (that is, $p - x$ is sufficiently small, or $f(x)$ is sufficiently small), return x and stop iterating;

(4)   Examine the sign of $f(x)$ and replace either $(p, f(p))$ or $(q, f(q))$ with $(x, f(x))$ so that there is a zero crossing within the new interval.

Thus the final formula we suggest and propose for the scale-down is:

$$\frac{1-\eta}{1-\eta_{ref}} = 0.50 + 0.50 \cdot \left[\frac{Re_{ref}}{Re} \right]^{\left[0.084 \left(\frac{Re_{ref}}{Re} \right)^{0.25} \right]} \tag{5}$$

The coefficients in Equation (5) provide a good agreement with the values reported in literature, particularly with those of Wiesner. Figure 3 reports the location of the points obtained by applying Equation (5) to several ultra-micro gas turbine devices designed by several independent research groups [10,12,15].

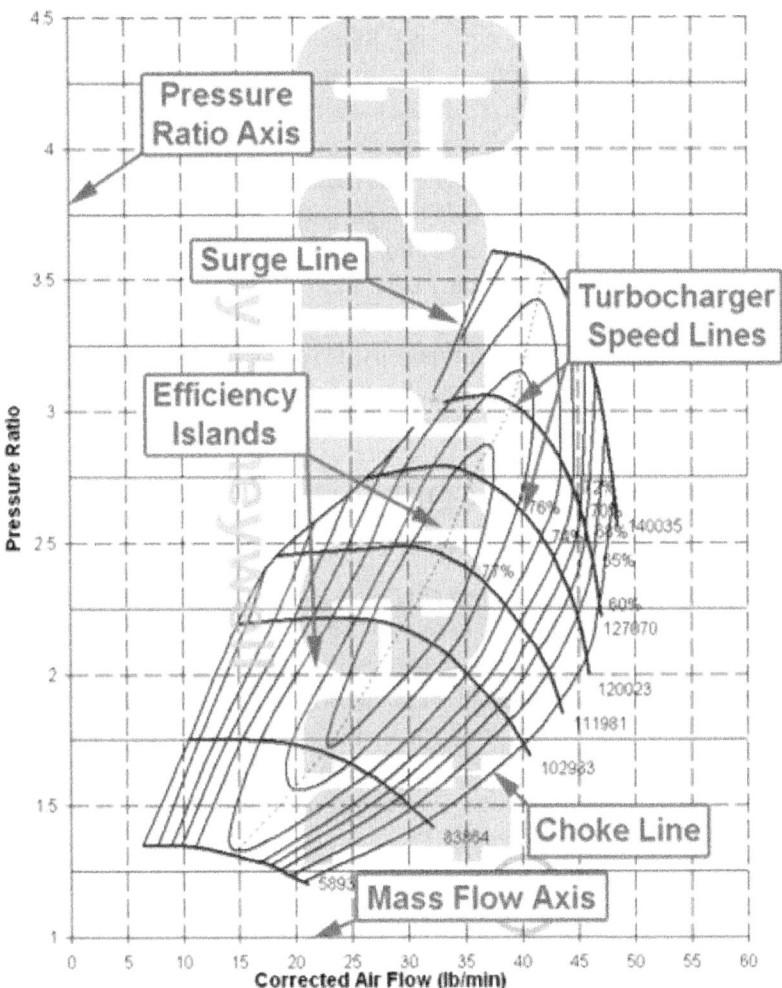

Figure 3: Typical Garrett® compressor maps.

RESULTS OF THE COMPUTATIONAL FLUID DYNAMICS VALIDATION

A series of CFD simulations was necessary to obtain the data needed to reconstruct the correlations presented here: more details are offered in previous articles by the same group [11,13,14,16], so that only some essential considerations will be reported here. The analysis compared the collected data to the resulting efficiency predicted by the charts for the given n_s and d_s. Both correlations predict an efficiency of the 1:10 model remarkably different from the one predicted by the Balje charts (60.6% and 75% *versus* 78% and 88% for compressor and turbine respectively). Since the charts refer to well-designed, commercial scale configurations, this difference is likely to be related to both the machine geometry and size. On the basis of the data reported by Wiesner [2], in which different small-sized compressors with 2-D and 3-D geometries were compared, it can be concluded that—all other parameters being the same—the penalty brought about by a "2-D" stage with respect to a 3-D one is about 18%. This constitutes an indirect validation for the correlations derived by our numerical calculations. In fact, $0.75 \times 0.82 = 0.615$, which is very close to our 0.606, where 0.82 is the efficiency deduced by the Balje chart and 0.606 is the efficiency calculated by the simulations. It seems to indicate that the penalty due to a non-optimal geometry can be modeled by a function of the Reynolds number in which at least one of the coefficients is geometry dependent.

VALIDATION OF THE CORRELATION ON COMMERCIAL ULTRA-MICRO MACHINES

The goal of our study was to generate a procedure for preliminary design applications. What is proposed here is to use Equation (3) to scale the efficiency with the Reynolds number calculated as in Equation (2). The coefficients in Equation (3) were been calculated, as described above, by an accurate fitting of the results of a series of numerical simulations. It is necessary, therefore, to confirm such values by means of the analysis of existing compressors and turbines. Garrett data relative to several turbo machinery of different size have been used for this validation step. The comparison was carried out on the basis of the original operative maps, courtesy of Garrett (Figure 3 and Figure 4). Provided a machine of known geometry and with known operative maps is available (in our case, the Garrett GT12, C224/45 trim compressors and GT12, T229/72 trim turbine), the method adopted to collect and compare the available compressor data is the following:

(1)   Based on the compressor/turbine map of the selected machine, calculate Re with Equation (2);

(2)   With the design value for the rpm, calculate the specific speed to enter the map;

(3)   Draw on the map the curve for that specific speed "n_s", and extrapolate for every value of the flow rate, the compression/expansion ratio and the efficiency;

(4)   Calculate the corrected flow rate accordingly;

(5)   Use Equation (3) to calculate the reference efficiency η_{ref};

(6)   Use η_{ref} to calculate, with Equation (5), the polytropic efficiency η_{pol};

(7)   Calculate the polytropic work;

(8)   Finally, compute n_s and d_s.

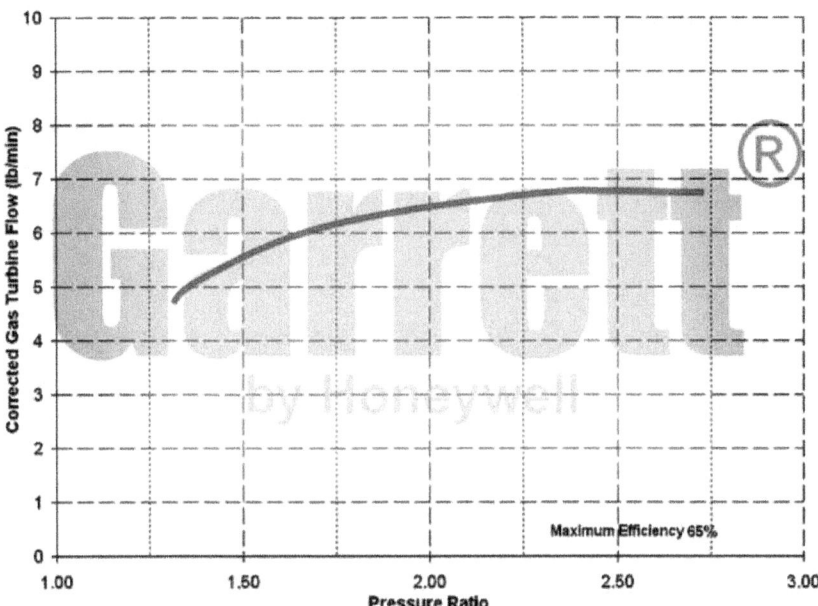

Figure 4: Typical Honeywell/Garrett® turbine maps. Reprinted with permission from [17].

Compressor Maps

Three different GT devices, all manufactured by the Garrett Corporation, have been considered in this work, identified by their impeller exit diameter, $D =$ 32, 34, 38 mm respectively. The graphs are reported in Figure 5. Following the suggestions of one of the reviewers, the validation was extended by use of additional maps by Mitsubishi & KKK. We compared some of the Mitsubishi maps (for the TD04-09B, -13G and -15G) and the KKK 04-14 (Figure 6) with the original Garrett maps and found that all of them cover a similar Re range and that the impellers sizes are very similar, which was to be expected (Figure 7).

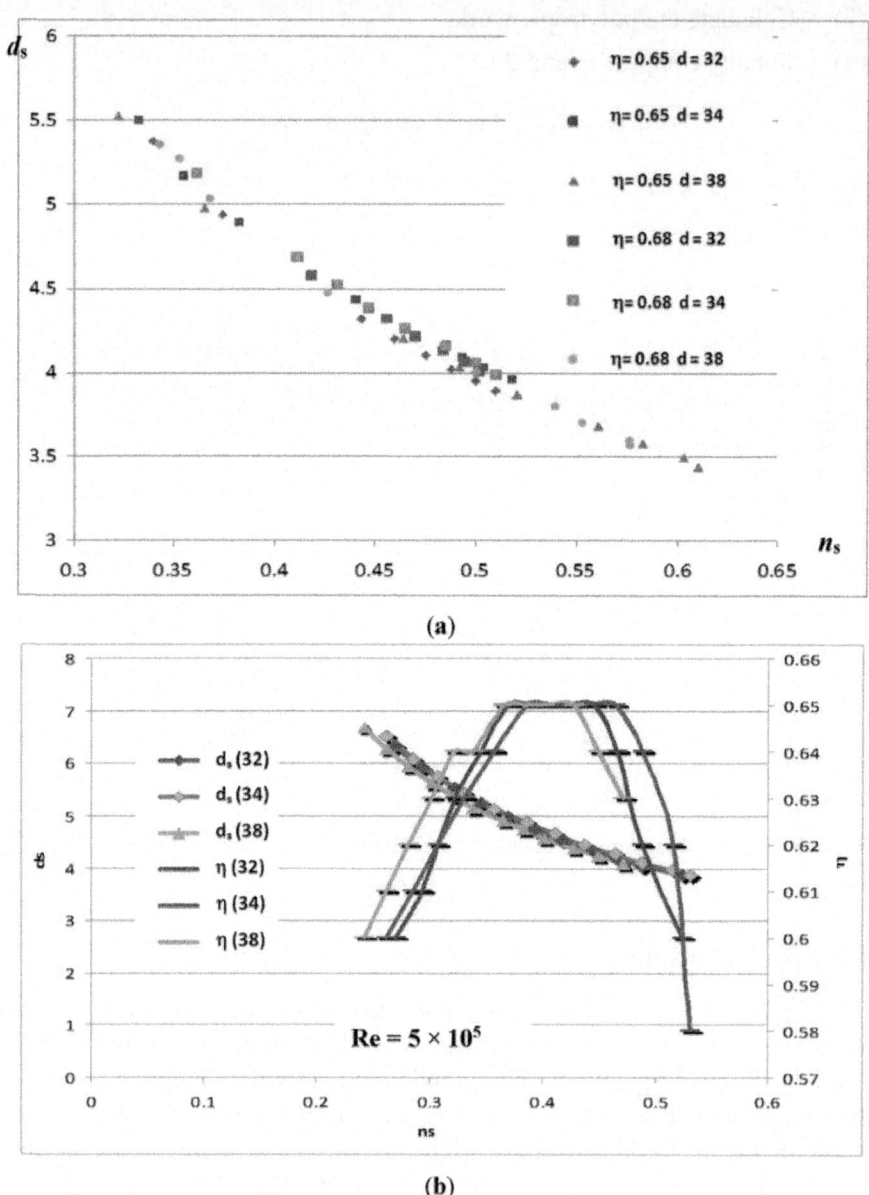

Figure 5: (a) Iso-efficiency points (Cordier line) for different compressor impeller diameter; and **(b)** d_s and efficiency curves for a fixed Re and different compressor impeller diameter.

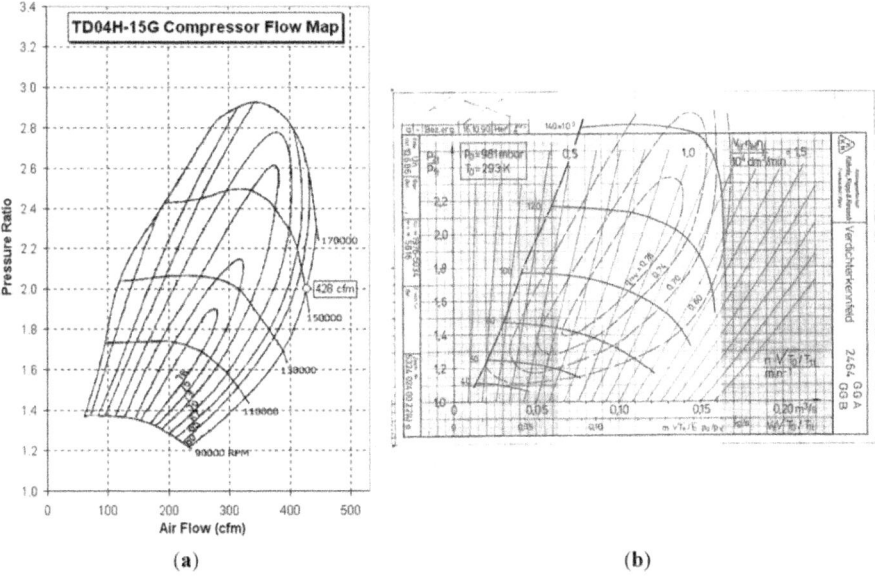

(a)   (b)

Figure 6: Compressor maps for (**a**) the Mitsubishi TD04 09B and (**b**) the KKK 04-14 used for the comparison and evaluation of proposed Equation (5).

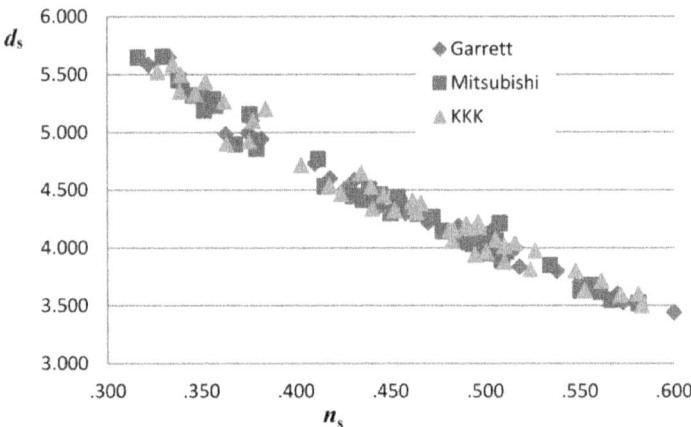

Figure 7: Comparison of the trends of the η_{max} of three different compressors.

The very good agreement between the predictions of Equation (5) and the available experimental data confirms the validity of the proposed efficiency scaling. Figure 5a shows the trend of the curves of iso-efficiency for three different diameters (32, 34 and 38 mm) of the compressor impeller, for several values of the Reynolds number. Here we only report two different values of the efficiency, corresponding to Re $= 5 \times 10^5$ and 7×10^5, and their dispersion for different values of ns at design rotational speed ($\omega = 15,708$ rad/s, or 150,000 rpm). Figure 5b, relative to a single Re value (in the validation process, these considerations have been repeated for every value of Re), shows the efficiency curves corrected according to Equation (5) and the d_s curves as functions of n_s. Figure 8 illustrates "the corrected" optimal curves in the n_s/d_s plane for three different values of Re: these curves were calculated with Equation (5). The same procedure was repeated for the remaining compressor maps. Figure 9 shows the values obtained using the Mitsubishi and KKK compressor maps respectively. Different diameters have been considered, as well as different Reynolds numbers, while the design angular velocity remains constant at $\omega = 15,708$ rad/s. The results obtained using the proposed formula show a good correspondence with the experimental values: in our judgement, this is a direct consequence of the philosophy of our approach, which applies the scaling only to the Re depending portion of the correction factor. The factor a does not change in Equation (3), so that the "inviscid" losses, which do not scale with Re, remain unchanged.

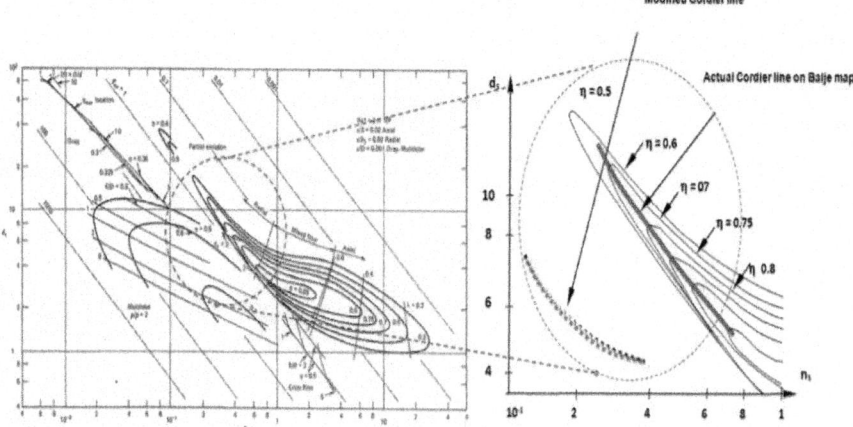

Figure 8: The low-Re Cordier curve on the standard Balje compressor map. Comparison between the actual Cordier line and the Modified Cordier line.

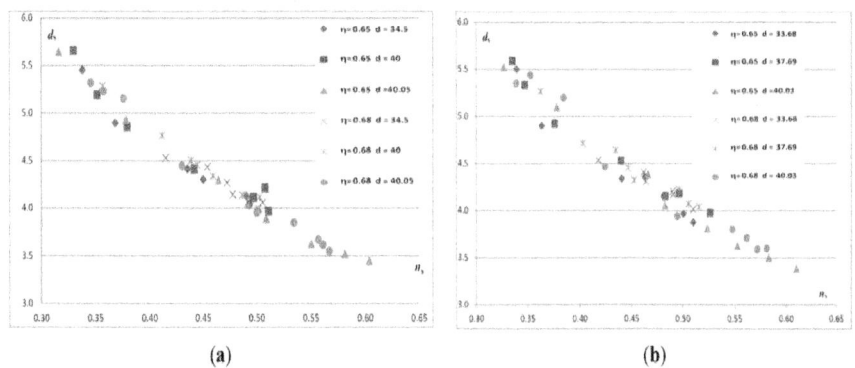

(a)   (b)

Figure 9: Iso-efficiency points (Cordier line) for different compressor impeller diameter: (**a**) Mitsubishi; and (**b**) KKK.

Turbine Maps

Repeating the same procedure for radial turbines, we again obtained a good agreement between the predictions of Equation (5) and the experimental data referring to several commercial turbines. As for the compressors, the calculated $\eta = f(n_s/d_s)$ curves lie in an area below the optimum and shifted to the left with respect to the peak values. These results are shown in Figure 10 only for Garrett turbines: they prove that—for a given turbine mass flow rate and turbine inlet temperature—either lower ω or higher specific work (*i.e.*, a different blade design) must be adopted to attain the "optimal" efficiency, which is anyhow significantly lower than the one predicted by the Balje chart. Similar results and conclusions have been obtained for the Mitsubishi and KKK turbines.

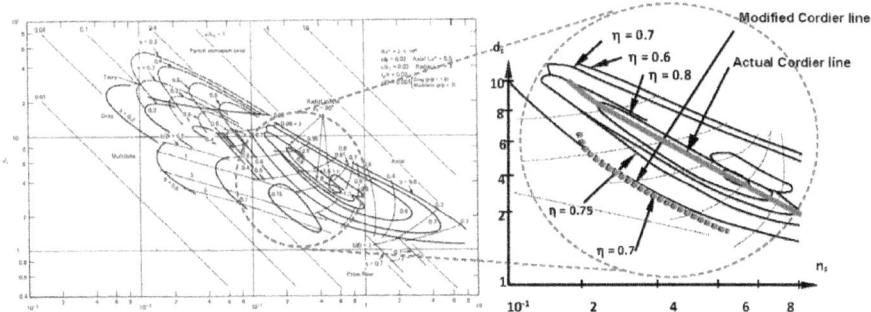

Figure 10: Radial turbines: the low-Re Cordier curve on the standard Balje turbine map. Comparison between the actual Cordier line and the Modified Cordier line.

THE PROPOSED OPERATIONAL APPROACH FOR THE SCALING-DOWN OF THE BALJE MAPS

As previously stated, we assumed as a first approximation that the manufacturing process and clearances are compatible with the parameters in Balje charts: this is a necessary step to ensure congruent results. Next, we neglect relative roughness effects: this is not strictly necessary within the present framework but it is essential for simplifying the treatment discussed here so that attention is placed solely on the influence of the Reynolds number (as derived from numerical calculations). Since the newly derived $\eta = f(Re)$ correlation has been shown: (a) to be in reasonable accord with the numerical and experimental results; and (b) to lead to the Balje limit for $Re = Re_{Balje}$, the following procedure can be proposed to extend into the ultra-micro scale the preliminary design selection at larger scales:

(1)   On the basis of the actual design data, identify an optimal value of n_s/d_s for radial compressors/turbines on the Balje chart;

(2)   Calculate the Re of the micro-scale machine;

(3)   Introduce an efficiency factor <1 given by Equation (5) that multiplies the total-to-static efficiency extracted from the chart;

(4)   Calculate the Euler work from the design data and extract the angular velocity ω and the impeller diameter D from n_s and d_s;

(5)   Check that $U < U_{max}$ (the allowable material stress limit for ultra-micro components may be higher): if this constraint is not abided by, choose a different n_s/d_s pair and go back to Step (1);

(6)   If the computed Re is different from the value assumed in Step (3), go back to Step (3), substitute the old Re number with the new one, and iterate.

CONCLUSIONS

The paper suggests a simple method to scale the Balje maps for low Re values. The method represents a generalization and a systematization of preliminary studies previously published on this topic by the authors' group. It is valid for adiabatic transformations, and thus its results can be directly compared to the classical Balje charts. Though the efficiency correction as a function of Re is completely similar to—and congruent with—the classical Stodola and Wiesner method, its novelty resides in a more accurate calculation of the correlation coefficients, which are obtained by a combination of numerical and laboratory experiments.

In spite of the relatively small number of CFD simulations, the results display a remarkably good agreement with the real operational map data of the devices object of the present study. This reinforces the belief in the correctness of the proposed preliminary design procedure for application to the small Re, ultra-micro scale range.

In particular, the results agree well with the available operational maps of a broad series of commercially available compressors and turbines. The most important practical result is the possibility of making a direct use of the Balje maps in the small-scale range by considering a Cordier line shifted towards lower n_s and d_s (or, for the same n_s and d_s, towards lower efficiencies). The Re-dependence of the results is apparent, and Equation (5) constitutes a good and reliable starting point for a preliminary design procedure, which in followed in any case in real industrial cases by prototype testing.

ACKNOWLEDGMENTS

The research has been partially sponsored by the Italian Army HQ within the framework of PNR 714.

AUTHOR CONTRIBUTIONS

Roberto Capata's contribution dealt with the experimental tests and comparisons, while Enrico Sciubba was in charge of all aspects related to turbomachinery and CFD simulations.

REFERENCES

1. Gong, X.; Chen, R. Total pressure loss mechanism of centrifugal compressors. Mech. Eng. Res.2014, 4, 45–59.

2. Wiesner, F.J. A new appraisal of Reynolds number effects on centrifugal compressor performance. J. Eng. Gas Turbines Power 1979, 101, 384–396.

3. Stodola, A.; Lowestein, C. Steam and Gas Turbines; McGraw-Hill: New York, NY, USA, 1927; Volume 1.

4. Balje, O. Turbomachines; John Wiley & Sons: New York, NY, USA, 1981.

5. Dixon, S.L. Fluid Mechanics and Thermodynamics of Turbomachinery, 3rd ed.; Pergamon Press: Exeter, UK, 1992; pp. 1–263.

6. Japikse, D. Centrifugal Compressor Design and Performance; Concept ETI Inc.: White River Junction, VT, USA, 1996.

7.   Japikse, D.; Baines, N.C. Introduction to Turbomachinery; Concept ETI Inc.: White River Junction, VT, USA, 1997.

8.   Casey, M.V. The effects of Reynolds number on the efficiency of centrifugal compressor stages.J. Eng. Gas Turbines Power 1985, 107, 541–548.

9.   Strub, R.A.; Bonciani, L.; Borer, C.J.; Casey, M.V.; Cole, S.L.; Cook, B.B.; Kotzur, J.; Simon, H.; Strite, M.A. Influence of the Reynolds number on the performance of centrifugal compressor. J. Turbomach. 1987, 109, 541–544.

10.   Peirs, J.; Reynaerts, D.; Verplaetsen, F. A Microturbine for Electric Power Generation. J. Sensors Actuators A Phys. 2004, 113, 86–93.

11.   Capata, R.; Sciubba, E.; Silva Caaveiro, G. Effects of the Reynolds Number on the Efficiency of an Ultra-micro Compressor. In Proceedings of the ECOS 2007, Padova, Italy, 25–28 June 2007.

12.   Ishihama, M.; Sakai, Y.; Matsuzuki, K.; Hikone, T. Structural Analysis of Rotating Parts of an Ultra Micro Gas Turbine. In Proceedings of the International Gas Turbine Congress, Tokyo, Japan, 2–7 November 2003.

13.   Capata, R.; Sciubba, E. Use of Modified Balje Maps in the Design of Low Reynolds Number Turbo Compressors. In Proceedings of the ASME 2012 International Mechanical Engineering Congress and Exposition-IMECE Conference, Houston, TX, USA, 9–15 November 2012; pp. 835–841.

14.   Silva Caaveiro, G. Micro Gas Turbines Numerical Analysis. Master's Thesis, University of Roma "Sapienza", Roma, Italy, June 2007.

15.   Epstein, A.H. Millimetre-Scale, MEMS Gas Turbine Engines. In Proceedings of the ASME Turbo Expo 2003, Atlanta, GA, USA, 16–19 June 2003.

16.   Iandoli, C.L.; Sciubba, E. 3-D numerical calculation of the local entropy generation rates in a radial compressor stage. Int. J. Thermodyn. 2005, 8, 83–94.

17.   Turbocharger Specs. Garrett GT20-GT2052-52 TRIM-225HP. Available online: http://turbochargerspecs.blogspot.it/2011/02/garrett-gt20-gt2052-225-hp.html (accessed on 15 May 2015).

Chapter 7

NUMERICAL ANALYSIS OF HORIZONTAL-AXIS WIND TURBINE CHARACTERISTICS IN YAWED CONDITIONS

Masami Suzuki

Department of Mechanical Systems Engineering, University of the Ryukyus, Okinawa, Japan

ABSTRACT

Computational fluid dynamics (CFD) modeling and experiments have both advantages and disadvantages. Doing both can be complementary, and we can expect more effective understanding of the phenomenon. It is useful to utilize CFD as an efficient tool for the turbomachinery and can complement uncertain experimental results. However the CFD simulation takes a long time for a design in generally. It is need to reduce the calculation time for many design conditions. In this paper, it is attempted to obtain the more accurate characteristics of a wind turbine in yawed flow conditions for a short time, using a few grid points. It is discussed for the reliability of the experimental results and the CFD results.

INTRODUCTION

Computational fluid dynamics (CFD) modeling and experiments have both advantages and disadvantages. Doing both can be complementary, and we can expect more effective understanding of the phenomenon. Although CFD has more advantages than experiments for the prediction where experiments are difficult to carry out, generally when compared with experimental results, it is difficult to obtain reliable results for a large domain by using CFD. However, it is possible to obtain useful CFD results based on verification by the experimental results. Moreover, experiments cannot deliver correct results for any arbitrary condition due to limitations to experimental equipment, measurement errors and problems with measurement systems. It is useful to utilize CFD as an efficient tool for the turbomachinery and can complement uncertain experimental results. However the CFD simulation takes a long calculation time for a design in generally. It is need to reduce the calculation time for many design conditions. In this paper, it is attempted to solve the more accurate characteristics of a wind turbine for a short time even a personal computer, using coarse grid [1]. In this paper the wind turbine characteristics of the yawed condition are discussed including the reliability of the experimental results and the CFD results.

NUMERICAL METHOD

The in-house code used is an incompressible finite volume Navier-Stokes solver which is developed originally. The solver is based on structured grids and the use of curve-linear boundary fitted coordinates. The grid arrangement is collocated (Perić et al. [2]) and the Rhie and Chow interpolation method [3] is used. The SIMPLE algorithm (Patankar [4]) is used for pressure-velocity coupling. The convection term is calculated using the QUICK scheme (Leonard [5]) and the other terms in space are calculated using the 2nd order difference schemes. It is well known that sophisticated turbulence models do not always produce better results than the very simple models. For practical applications that are computationally expensive it is often wiser to use a simple approach. Therefore the proven and computationally efficient Launder-Sharma low-Reynolds-number k-e turbulence model [6] is used in this report.

WIND TURBINE AND AERODYNAMIC FORCE ACTING TO BLADE

Figure 1 shows a schematic view of experimental apparatus for a wind turbine carried out by Vermeer [7]. A two bladed wind turbine is situated in front of the wind turbine. The wind turbine has diameter of 1.2 m and the blades consist of NACA 0012 airfoil and the chord length of 0.08 m. The experiment is conducted at wind velocity of 5 m/s, and the measured data are the wind velocity, the number of rotation, the torque, and the thrust. Moreover, Haans et al. [8,9] measure with the same experiment equipment about the thrust according to a yawed wind (0°, ±15°, ±30°, ±45°) of 5.5 m/s in speed and observe the tip vortex by flow visualization.

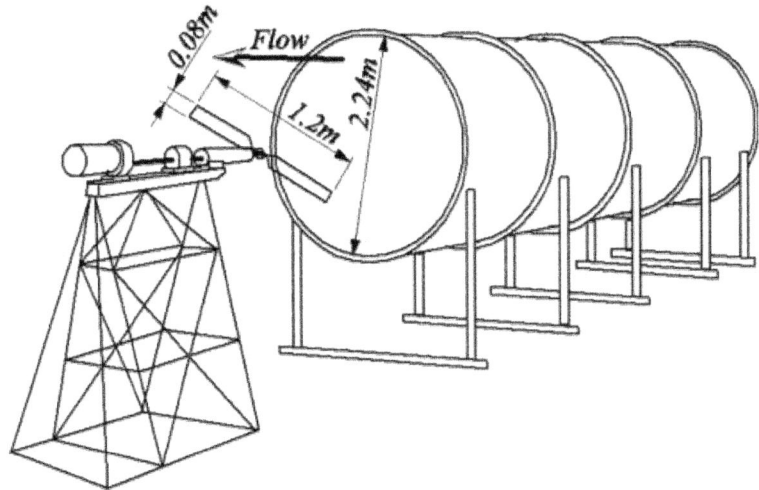

Figure 1: Schematic view of experimental apparatus of delft university of technology (Vermeer [7]).

Figure 2 shows the relation with the fluid force acting to the blade element of the wind turbine with radius, r, the angle of pitch, q, the angle of attack, a, the lift, L, the drag, D, the tangential force, F_t, the axial force, F_a, i.e. thrust force, the axial velocity, V_a, the tangential velocity, rw, the rotational speed, w, and the relative velocity, W.

Figure 3 shows the pitch angle, q, at each radius position, r. The pitch angle is changed linearly from q_{tip} + 4° to q_{tip} between nondimensional radius r/R = 0.3 and 0.9, and it is fixed to the pitch angle at tip between r/R = 0.9 and 1.0. The calculation is performed by a pitch angle at tip, that is, q_{tip} = 2°. Relation among the lift, L, the drag, D, the tangential force, F_t, and the axial force, F_a, inFigure 2 are written by:

$$F_t = L\sin(\theta+\alpha)-D\cos(\theta+\alpha)$$
$$F_a = L\cos(\theta+\alpha)+D\sin(\theta+\alpha) \qquad (1)$$

Under the no stall conditions which are small angle of attack, Equation (1) is approximated as follows:

$$F_t \cong L(\theta+\alpha)-D$$
$$F_a \cong L \qquad (2)$$

Since the axial force, F_a, i.e. thrust force, is predicted by the same accuracy as lift. On the other hand, since the tangential force, F_t, i.e., torque, is strongly influenced of drag, D, and it serves as the difference of the force by the lift and the drag, the produced force becomes small. For this reason, the predicted accuracy of torque is reduced than one of the thrust force.

Figure 4 shows the arrangement of yaw and azimuth angle.

The Reynolds number $Re = UR/v$ is expressed by the turbine radius, R, the wind velocity, U, and the kinematic viscosity of air, v. The characteristics of wind turbine are expressed by the tip speed ratio, l, the power coefficient, C_p, and the thrust coefficient, C_a,

$$\lambda = \frac{R\omega}{U}, C_P = \frac{T\omega}{\frac{1}{2}\rho U^3 \pi R^2}, \quad C_a = \frac{F_a}{\frac{1}{2}\rho U^2 \pi R^2}$$

(3)

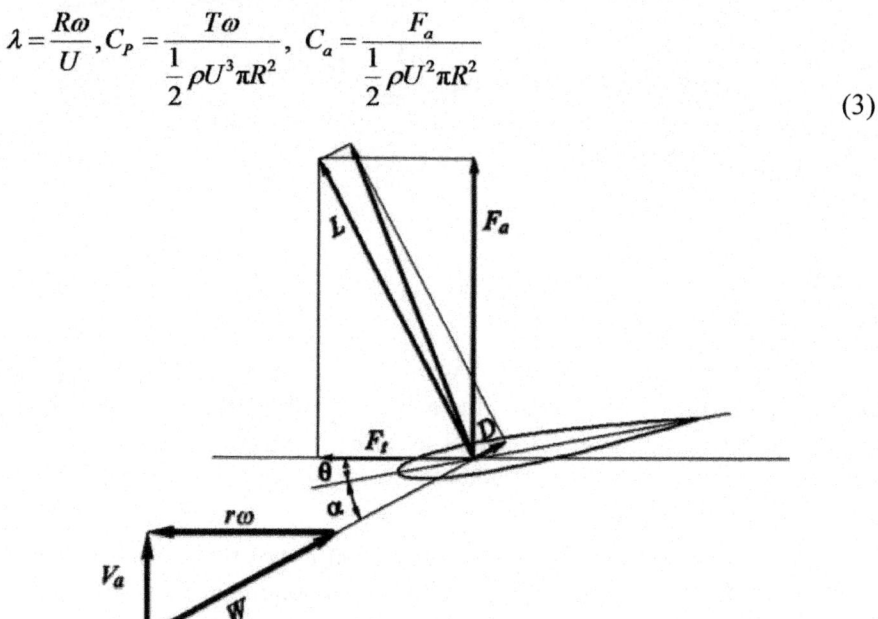

Figure 2: Fluid force acting a blade.

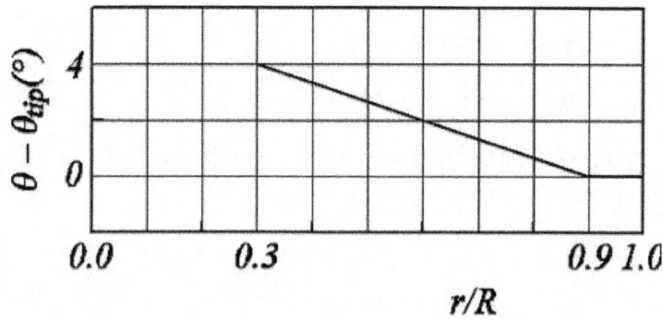

Figure 3: Pitch angle for radius of blade.

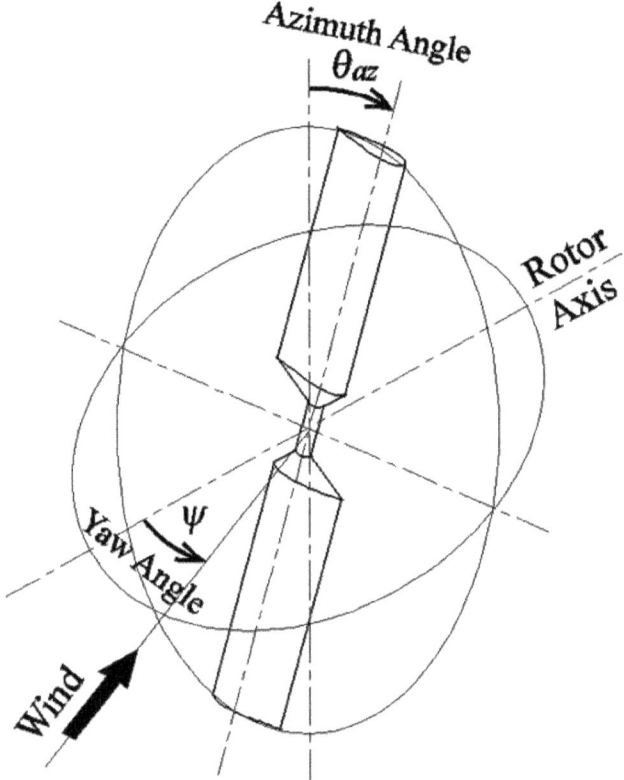

Figure 4: Arrangement of yaw and azimuth angle.

where the tip speed, U_{tip}, the wind velocity, V_a, the torque, T, the thrust, F_a, the air density, r.

COMPUTATIONAL GRID

Figure 5 shows the upper half domain of the computational grid around the wind turbine rotor which is a sphere domain. The radius of the sphere is made into ten times of the rotor radius, and the external boundary is located at the 75 times of the chord length from the rotor axis. In addition, the internal diameter of the wind tunnel is about twice of wind turbine diameter, as shown in Figure 1, and the computational domain is sufficient wider than the experiment condition, that is 5 times of the internal diameter of a wind tunnel. The O-O type grid is enabling a suitable grid arrangement, being able to arrange many grid points along wing surface without

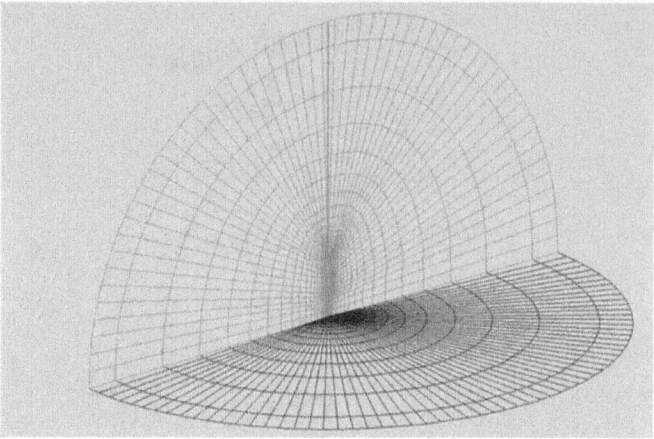

Figure 5: 3-D computational O-O grid around blade of wind turbine.

distributing many points to unnecessary parts. The number of grid points is 130 around the configuration, 56 points in the spanwise, 57 points normal to the surface direction, and the 414,960 points in total. The grid is generated using a algebraic grid generation method (Eriksson [10]) based on the transfinite interpolation method which gives 5×10^{-5} in a direction normal to the nearwall grid spacing to unit rotor radius, and y^+ values of less than 0.65.

COMPARISON BETWEEN COMPUTATION AND EXPERIMENT IN NON-YAWED FLOW

Figure 6 shows the power coefficient of $q_{tip} = 2°$. The experimental results show the characteristics that the maximum power coefficient, C_p appears at $l = 7.5$, the stall region appears below $l = 6$, and the power coefficient decreases from above $l = 7.5$, because the angle of attack becomes smaller as increase of tip speed ratio. The power coefficients are in good agreement with experimental results. The big difference between the computational and the experimental results produces near just after stall angle where the tip speed ratio is $l = 6 \sim 3$ [1]. Since the present turbulence models cannot fully predict the transition from laminar to turbulent flow, and cause to delay stall. About this, we will expect for development of a future turbulence model. Since the leading edge separation is completely occurred in the region less than $l = 3$ in tip speed ratio, it is comparatively easy to catch also in the CFD, and the results become nearly equal to ones of the experiment. For the region is larger than $l = 6.75$ in tip speed ratio which becomes small angle of attack not to stall, the characteristics can fully be predicted by the numerical computation.

Figure 7 shows the thrust coefficient. The computational results of a thrust coefficient, C_a agree well with the experimental results all over the region, because the thrust is dominated by the lift and little influence of a drag. Furthermore, the lift coefficient can expect accurate

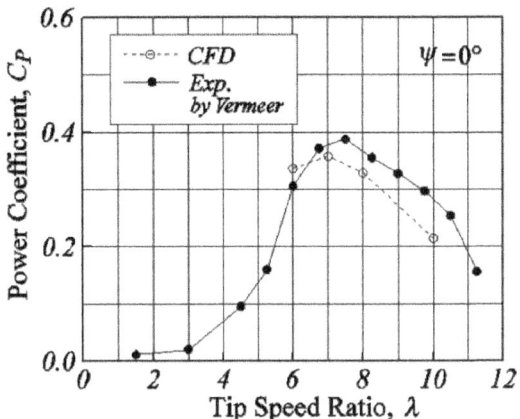

Figure 6: Power coefficients of non-yawed condition.

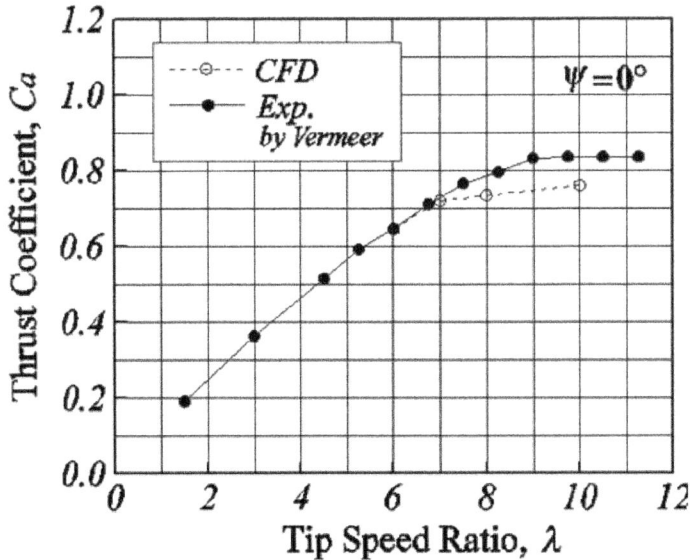

Figure 7: Thrust coefficients of non-yawed condition.

prediction for the CFD, because even the potential calculation is as small different as about 10% from the experimental result.

The thrust force is nearly equal to the lift, while the tangential force is strongly influenced of a drag, because it becomes the difference of L (a + q) and D from Equation (2), it turns into the small force of less than 10% as compared with a thrust force. For this reason, it is easy not only in numerical computation but also in the experiment to produce a big error for the tangential force.

WIND TURBINE CHARACTERISTICS IN YAWED FLOW

Haans et al. [8] measure only the thrust coefficient as experimental data, and the measurement about the power coefficient has not gone. For this reason, only thrust coefficients are compared with experimental results here. And assuming that only the turbine axis component of wind velocity influences the characteristics of the yawed flow, the simple wind turbine characteristics computed using the experimental data of $\psi = 0°$ are also described together, that is, as characteristics of yawed flow:

$$\lambda(\psi) = \frac{R\omega}{U} = \lambda(0)\cos\psi,$$

$$C_P(\lambda(\psi),\psi) = \frac{T\omega}{\frac{1}{2}\rho U^3 \pi R^2} \cong C_P(\lambda(0),0)\cos^3\psi,$$

$$C_a(\lambda(\psi),\psi) = \frac{F_a}{\frac{1}{2}\rho U^2 \pi R^2} \cong C_a(\lambda(0),0)\cos^2\psi$$

$$(4)$$

are introduced as the estimation equation, where $C_P(l, 0)$ and $C_a(l, 0)$ are the power coefficient and thrust coefficient for the tip speed ratio at yaw angle of 0°, respectively. In relation to this, Maeda et al. [11] have reported that the maximum power coefficients becomes in the non-yawed value times between $\cos^2\psi$ and $\cos^3\psi$ by the experiment.

Figures 8-10 show the thrust coefficients for $\psi = \pm15°, \pm30°$ and $\pm45°$. First, the computational fluid dynamics (CFD) results are compared with the simple prediction results (Cal.) calculated by Equation (4) using the experimental data of yawed angle $\psi = 0°$ by Vermeer [6]. The results of Equation (4) are well in agreement with CFD results for $\psi = 15°, 30°$ and 45°. For tip speed ratio, $l = 8$ and 10, although the difference in about ten percent has appeared, it is because the CFD results have the difference from the experiment data in the yawed angle 0°. Thus, the result of the simple presumed formula and

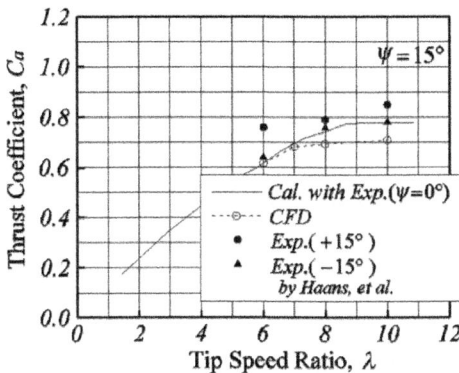

Figure 8: Thrust coefficients of 15° yawed angle.

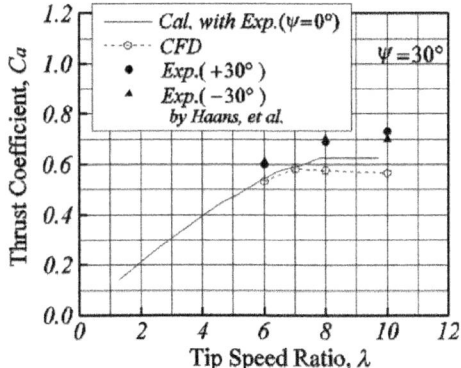

Figure 9: Thrust coefficients of 30° yawed angle.

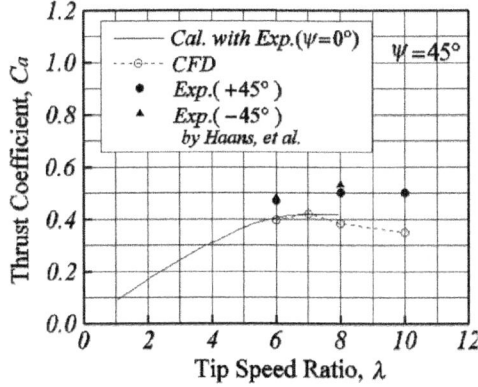

Figure 10: Thrust coefficients of 45° yawed angle.

CFD corresponds well, it is suggested that the cross flow along a surface of revolution does not influence to the thrust.

On the other hand, although the experimental results of $\psi = 15°$ are well in agreement with a simple formula and the CFD, the experimental results become bigger value than the value predicted by CFD as large yaw angle such as $\psi = 30°$ and $45°$. But the big error is also included in the experimental results of $\psi = 15°$ shown in Figure 5. Essentially, since the flow becomes symmetrical for the positive/negative of yaw angle, the thrust coefficient must be in agreement. However the difference between positive and negative yaw angle appears 10% - 20% in the experimental results. Haans et al. [8] shows judgment because the velocity distribution of the wind tunnel exit does not become uniform. Thus, the uncertain element is also contained in the experimental result and we want to consider as the future work about the difference between the experiment and the CFD result.

Figures 11-13 show the power coefficient for the yaw angles $\psi = 15°$, $30°$ and $45°$. The CFD results are compared with the simple prediction results (Cal.) calculated by Equation (4) using the experimental data of yaw angle $\psi = 0°$. Although the CFD results of $\psi = 0°$ is smaller, about 0.1, than the experimental result of power coefficient, the tendency is well in agreement for the tip speed ratio. About the yaw angle, $\psi = 15°$, $30°$, $45°$, the results of a simple formula (4) and the results of CFD show good coincidence, and it is shown that the macroscopic amount of time averages like the thrust coefficient and the power coefficient can express the characteristic by the comparatively easy relation.

FLUCTUATION OF THRUST AND POWER COEFFICIENT DEPENDING ON AZIMUTH ANGLE

As described by the previous section, although it is thought that the time average power and thrust are predicted comparatively well in the simple formula (4), which is calculated using the characteristic of the nonyaw angle, in order to predict the fatigue load inflicted to wings, it is important to presume correctly the unsteady characteristic for the azimuth angle.

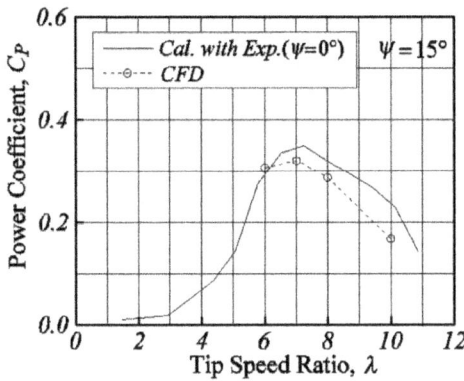

Figure 11: Power coefficients of 15° yaw angle.

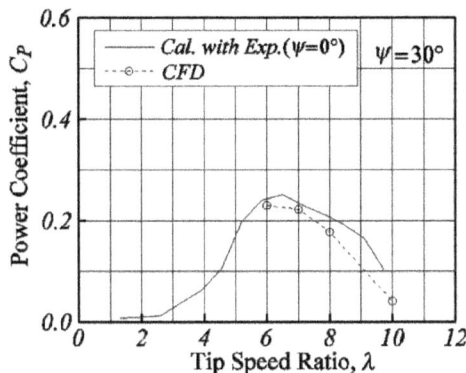

Figure 12: Power coefficients of 30° yaw angle.

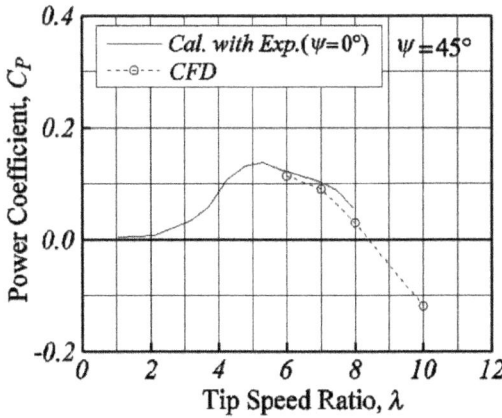

Figure 13: Power coefficients of 45° yaw angle.

Figures 14-16 show the fluctuation of thrust force, C_a, for azimuth angle, θ_{az}. The range of thrust fluctuation for the azimuth angle increases in connection with the increase in the yaw angle, ψ, or the tip speed ratio, l, and the range of fluctuation has reached about 30% of the average value at l = 10 and ψ = 45°. The thrust force is the maximum at θ_{az} = 90° and the minimum at θ_{az} = 250° for the tip speed ratio, l = 6, the maximum at θ_{az} = 110° and the minimum at θ_{az} = 30° for l = 8, and the maximum at θ_{az} = 180° and the minimum at θ_{az} = 0° for l = 10. The thrust force is smoothed by the two blades to be canceled with each other blade.

Figures 17-19 show the fluctuation of power for the azimuth angle. The amplitude of the power fluctuation also increases in connection with the increase in the yaw angle or the tip speed ratio, the range of fluctuation has reached the twice of average value for the yaw angle of ψ = 45°. The power is the maximum at θ_{az} = 140° and the minimum at θ_{az} = 0° for l = 6, the maximum at θ_{az} = 170° and the minimum at θ_{az} = 5° for l = 8, and the maximum at θ_{az} = 160° and the minimum at θ_{az} = 350° for l = 10. The power of rotor is smoothed by the two blades to be canceled with each other blade. These show that the phase of the fluctuated waveform of the thrust force and the power tends to delay with the tip speed ratio.

CONCLUSIONS

While the CFD and the simple presumed formula are performed to grasp detailedly the wind turbine characteristics complementing the experimental results, the verification is performed by carrying out comparison and examination each other. The knowledge about the wind turbine characteristics of the yaw flow obtained by this study are shown as follows:

1) The simple calculation formula was introduced to presume the characteristics of the yaw flow.

2) The characteristics of the yaw flow are computed by the CFD, of which results are compared with the experimental results or the proposed simple calculation. It is

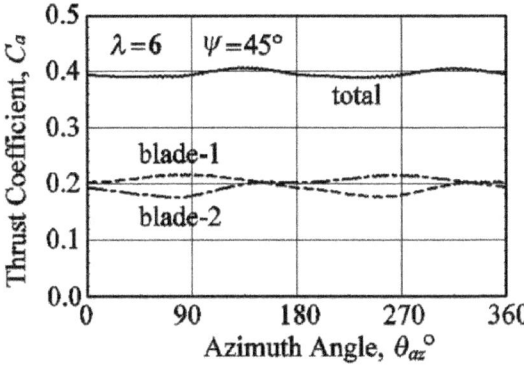

Figure 14: Fluctuation of thrust coefficients where the tip speed ratio is 6 and the yaw angle is 45°.

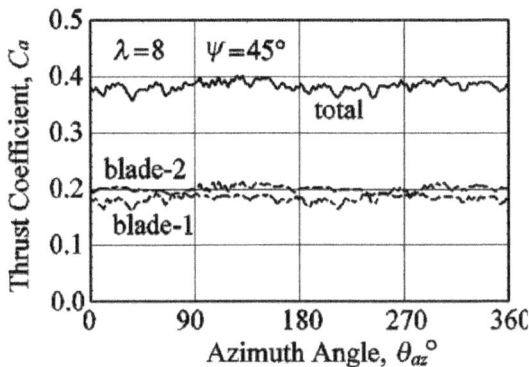

Figure 15: Fluctuation of thrust coefficients where the tip speed ratio is 8 and the yaw angle is 45°.

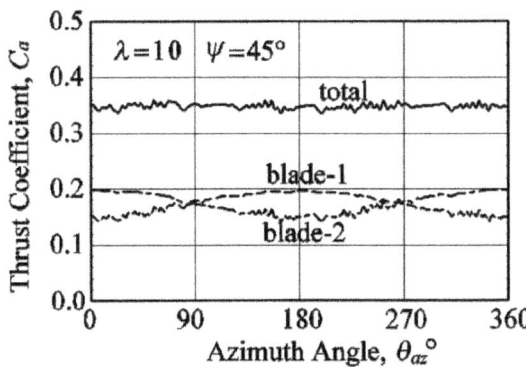

Figure 16: Fluctuation of thrust coefficients where the tip speed ratio is 10 and the yaw angle is 45°.

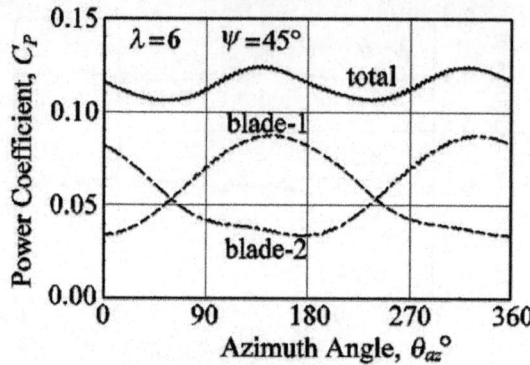

Figure 17: Fluctuation of power coefficients where the tip speed ratio is 6 and the yaw angle is 45°.

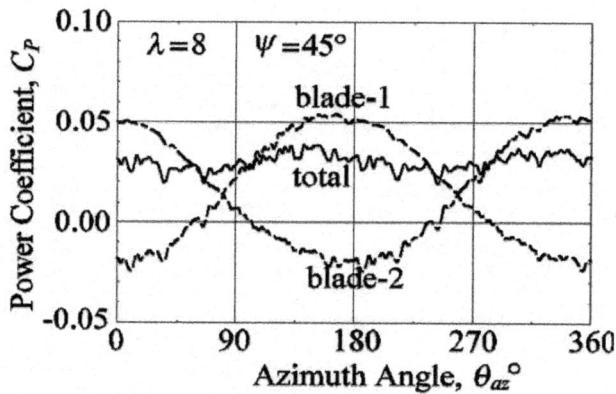

Figure 18: Fluctuation of power coefficients where the tip speed ratio is 8 and the yaw angle is 45°.

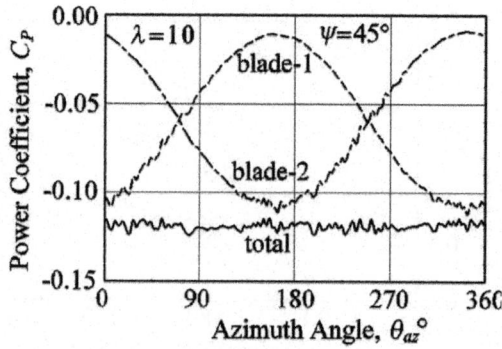

Figure 19: Fluctuation of power coefficients where the tip speed ratio is 10 and the yaw angle is 45°.

expected that the influence of yawed flow can be estimated roughly by the simple presumed formula, because the results of the simple presumed formula are in agreement with the CFD. However, since the CFD and the simple estimation are in the tendency to appear smaller about 10% - 20% than the experimental results of thrust coefficient, the validation of accuracy including also experiment will be required after this. Moreover, the power coefficients show also the similar tendency between the simple formula and the CFD results.

3)   The fluctuating torque and thrust force in one revolution affect the fatigue load of the blades. Although the total values of two blades are made smooth and the fluctuation decreases, the fluctuation of one blade is large. For example, in the yaw angle 45° and the tip speed ratio 10, the fluctuation ranges of the thrust coefficient and the torque coefficient reaches 3 times and twice of average value, respectively. Moreover, the tendency that the azimuth angle for which the thrust and the torque coefficient become the maximum and the minimum are delayed with the increase in the tip speed ratio appears.

REFERENCES

1.   M. Suzuki, "Evaluation of Experimental Results for Wind Turbine Characteristics by CFD," Proceedings of the 9th International Symposium on Experimental and Computational Aero-thermodynamics of Internal Flows (ISAIF9), Gyeongju, 2009, Paper No. 1D-2.

2.   M. Perić, R. Kessier and G. Scheuerer, "Comparison of Finite-Volume Numerical Methods with Staggered and Collocated Grids," Computers & Fluids, Vol. 16, No. 4, 1988, pp. 389-403. doi:10.1016/0045-7930(88)90024-2

3.   C. M. Rhie and W. L. Chow, "Numerical Study of the Turbulent Flow Past an Airfoil with Trailing Edge Separation," AIAA Journal, Vol. 21, No. 11, 1983, pp. 1525- 1532.doi:10.2514/3.8284

4.   S. V. Patankar, "Numerical Heat Transfer and Fluid Flow," McGraw-Hill, New York, 1980.

5.   B. P. Leonard, "A Stable and Accurate Convective Modeling Procedure Based on Quadratic Upstream Interpolation," Computer Methods in Applied Mechanics and Engineering, Vol. 19, No. 1, 1979, pp. 59-98. doi:10.1016/0045-7825(79)90034-3

6.   B. E. Launder and B. I. Sharma, "Application of the Energy-Dissipation Model of Turbulence to the Calculation of Flow near a

Spinning Disk," Letters in Heat Mass Transfer, Vol. 1, 1974, pp. 131-138. doi:10.1016/0094-4548(74)90150-7

7.   N. J. Vermeer, "Performance measurements on a Rotor Model with Mie-Vanes in the Delft Open Jet Tunnel," Institute for Wind Energy, Delft University of Technology, Delft, 1991, IW-91048R.

8.   W. Haans, T. Sant, G. van Kuik and G. van Bussel, "Measurement of Tip Vortex Paths in the Wake of a HAWT Under Yawed Flow Conditions," Journal of Solar Energy Engineering, Vol. 127, No. 4, 2005, pp. 456-463. doi:10.1115/1.2037092

9.   W. Haans, T. Sant, G. van Kuik and G. van Bussel, "Stall in Yawed Flow Conditions: A Correlation of Blade Element Momentum Predictions with Experiments," Journal of Solar Energy Engineering, Vol. 128, No. 4, 2006, pp. 472-480. doi:10.1115/1.2349545

10.   L. E. Eriksson, "Generation of Boundary Conforming Grids around Wing-Body Configurations Using Transfinite Interpolations," AIAA Journal, Vol. 20, No. 10, 1982, pp. 1313-1320. doi:10.2514/3.7980

11.   T. Maeda, Y. Kamada, J. Suzuki and H. Fujioka, "Rotor Blade Sectional Performance under Yawed Inflow Conditions," Journal of Solar Energy Engineering, Vol. 130, No. 3, 2008, Article ID: 031018. doi:10.1115/1.2931514

Chapter 8

TRANSITION MODELLING FOR TURBOMACHINERY FLOWS

F. R. Menter[1] and R. B. Langtry[2]

[1]ANSYS GmbH Germany

[2]The Boeing Company,USA

INTRODUCTION

In the past few decades a significant amount of progress has been made in the development of reliable turbulence models that can accurately simulate a wide range of fully turbulent engineering flows. The efforts by different groups have resulted in a spectrum of models that can be used in many different applications, while balancing the accuracy requirements and the computational resources available to a CFD user. However, the important effect of laminar-turbulent transition is not included in the majority of today's engineering CFD simulations. The reason for this is that transition modelling does not offer the same wide spectrum of CFD-compatible model formulations that is currently available for turbulent flows, even though a large body of publications is available on the subject. There are several reasons for this unsatisfactory situation. The first is that transition occurs through different mechanisms in different applications. In aerodynamic flows, transition is typically the result of a flow instability (TollmienSchlichting waves or in the case of highly swept wings cross-flow instability), where the resulting exponential growth of two-dimensional waves eventually results in a non-linear break-down to turbulence. Transition occurring due to Tollmien-Schlichting waves is often referred to as natural transition [1]. In turbomachinery applications, the main transition mechanism is bypass transition [2] imposed on the boundary layer by high levels of turbulence in the freestream. The high freestream turbulence levels are for instance generated by upstream blade rows. Another important transition mechanism is separationinduced transition [3], where a laminar boundary layer separates under the influence of a pressure gradient and transition develops within the separated shear layer (which may or may not reattach). As well, a turbulent boundary layer can re-laminarize under the influence of a strong favorable pressure gradient [4]. While the importance of transition phenomena for aerodynamic and heat transfer simulations is widely accepted, it is difficult to include all of these effects in a single model. The second complication arises from the fact that conventional Reynolds averaged NavierStokes (RANS) procedures do not lend themselves easily to the description of transitional flows, where both linear and non-linear effects are relevant. RANS averaging eliminates the effects of linear disturbance growth and is therefore difficult to apply to the transition process. While methods based on the stability equations such as the en method of Smith & Gamberoni [5] and van Ingen [6] avoids this limitation, they are not compatible with general-purpose CFD methods as typically applied in complex geometries. The reason is that these methods require a priori knowledge of the geometry and the grid topology. In addition, they involve numerous non-local operations (e.g. tracking the disturbance growth along each streamline) that are difficult to implement into today's CFD methods [7]. This is not to argue against the stability approaches, as they are an essential part of the desired "spectrum" of transition models required for the vastly different application areas and accuracy requirements. However, much like in turbulence modeling, it is important to develop engineering models that can be applied in day-to-day operations by design engineers on complicated 3D geometries. It should be noted that at least for 2D flows, the efforts of various groups has resulted in a number of engineering design tools intended to model transition for very specific applications. The most notable efforts are those of Drela and Giles [8] who developed the XFOIL code which can be used for modeling transition on 2D airfoils and the MISES code of Youngren and Drela [9], which is used for modeling transition on 2D turbomachinery blade rows. Both of these codes use a viscous – inviscid coupling approach which allows the classical boundary layer formulation tools to be used. Transition prediction is accomplished using either an en method or an empirical correlation and both of these codes are used widely in their respective design communities. A 3D wing or blade design is performed by stacking the 2D profiles (with the basic assumption that span wise flow is negligible) to create the geometry at which point a 3D CFD analysis is preformed. Closer inspection shows that hardly any of the current transition models are CFD-compatible. Most formulations suffer from non-local operations that cannot be carried out (with reasonable effort) in general-purpose CFD codes. This is because modern CFD codes use mixed elements and massive parallel execution and do not provide the infrastructure for computing integral boundary layer parameters or allow the integration of quantities along the direction of external streamlines. Even if structured boundary layer grids are used (typically hexahedra), the codes are based on data structures for unstructured meshes. The information on a body-normal grid direction is therefore not easily available. In addition, most industrial CFD simulations are carried out on parallel computers using a domain decomposition methodology. This means in the most general case that boundary layers can be split and computed on different processors, prohibiting any search or integration algorithms. Consequently, the main requirements for a fully CFD-compatible transition model are:

- Allow the calibrated prediction of the onset and the length of transition
- Allow the inclusion of different transition mechanisms
- Be formulated locally (no search or line-integration operations)
- Avoid multiple solutions (same solution for initially laminar or turbulent boundary layer)
- Do not affect the underlying turbulence model in fully turbulent regimes
- Allow a robust integration down to the wall with similar convergence as the underlying turbulence model
- Be formulated independent of the coordinate system
- Applicable to three-dimensional boundary layers

Considering the main classes of engineering transition models (stability analysis, correlation based models, low-Re models) one finds that none of these methods can meet all of the above requirements. The only transition models that have historically been compatible with modern CFD methods are the low-Re models [10,11]. However, they typically suffer from a close interaction with the transition capability and the viscous sublayer modeling and this can prevent an independent calibration of both phenomena [12, 13]. At best, the low-Re models can only be expected to simulate bypass transition which is dominated by diffusion effects from the freestream. This is because the standard low-Re models rely exclusively on the ability of the wall damping terms to capture the effects of transition. Realistically, it would be very surprising if these models that were calibrated for viscous sublayer damping could faithfully reproduce the physics of transitional flows. It should be noted that there are several low-Re models where transition prediction was considered specifically during the model calibration [14, 15, 16]. However, these model formulations still exhibit a close connection between the sublayer behavior and the transition calibration. Re-calibration of one functionality also changes the performance of the other. It is therefore not possible to introduce additional experimental information without a substantial re-formulation of the entire model. The engineering alternative to low-Re transition models are empirical correlations such as those of [17, 18 and 19]. They typically correlate the transition momentum thickness Reynolds number to local freestream conditions such as the turbulence intensity and pressure gradient. These models are relatively easy to calibrate and are often sufficiently accurate to capture the major effects of transition. In addition, correlations can be developed for the different transition mechanisms, ranging from bypass to natural transition as well as crossflow instability or roughness. The main shortcoming of these models lies in their inherently non-local formulation. They typically require information on the integral thickness of the boundary layer and the state of the flow outside the boundary layer. While these models have been used successfully in special-purpose turbomachinery codes, the non-local operations involved with evaluating the boundary layer momentum thickness and determining the freestream conditions have precluded their implementation into generalpurpose CFD codes. Transition simulations based on linear stability analysis such as the en method are the lowest closure level available where the actual instability of the flow is simulated. In the simpler models described above, the physics is introduced through the calibration of the model constants. However, even the en method is not free from empiricism. This is because the transition n-factor is not universal and depends on the wind tunnel freestream/acoustic environment and also the smoothness of the test model surface. The main obstacle to the use of the en model is that the required infrastructure needed to apply the model is very complicated. The stability analysis is typically based on velocity profiles obtained from highly resolved boundary layer codes that must be coupled to the pressure distribution of a RANS CFD code [7]. The output of the boundary layer method is then transferred to a stability method, which then provides information back to the turbulence model in the RANS solver. The complexity of this set-up is mainly justified for special applications where the flow is designed to remain close to the stability limit for drag reduction, such as laminar wing design.

Large Eddy Simulation (LES) and Direct Numerical Simulations (DNS) are suitable tools for transition prediction [20], although the proper specification of the external disturbance level and structure poses substantial challenges. Unfortunately, these methods are far too costly for engineering applications. They are currently used mainly as research tools and substitutes for controlled experiments. Despite its complexity, transition should not be viewed as outside the range of RANS methods. In many applications, transition is enforced within a narrow area of the flow due to geometric features (e.g. steps or gaps), pressure gradients and/or flow separation. Even relatively simple models can capture these effects with sufficient engineering accuracy. The challenge to a proper engineering model is therefore mainly in the formulation of a model that can be implemented into a general RANS environment. In this chapter a novel approach to simulating laminar to turbulent transition is described that can be implemented into a general RANS environment. The central idea behind the new approach is that Van Driest and Blumer's [21] vorticity Reynolds number concept can be used to provide a link between the transition onset Reynolds number from an empirical correlation and the local boundary layer quantities. As a result the model avoids the need to integrate the boundary layer velocity profile in order to determine the onset of transition and this idea was first proposed by [22]. Recently another class of locally formulated transition models have been proposed. They are based on modelling the laminar kinetic energy which is present already upstream of the actual transition location. This information is then applied to trigger the actual transition process. Methods of this kind have been proposed e.g. by Walters and Cokljat [23] and Pacciani et al. [24]. While the argumentation behind the derivation of these models is rather different from the g-ReQ model, the mechanisms by which transition is triggered is very similar. The current chapter is largely based on Langtry and Menter [25]. More recent articles on model validation and development can be found in [26-28].

MODEL FORMULATION

Basic Concept

The current approach is based on combining experimental correlations with locally formulated transport equations. The essential quantity to trigger the transition process is the vorticity or alternatively the strain rate Reynolds number which is used in the present model is defined as follows:

$$Re_v = \frac{\rho y^2}{\mu}\left|\frac{\partial u}{\partial y}\right| = \frac{\rho y^2}{\mu}S$$

(1)

where y is the distance from the nearest wall, S is the shear strain rate, r is the density and m is the dynamic viscosity. The vorticity Reynolds number it is a local property and can be easily computed at each grid point in an unstructured, parallel Navier-Stokes code. A scaled profile of the vorticity Reynolds number is shown in Figure 1 for a Blasius boundary layer. The scaling is chosen in order to have a maximum of one inside the boundary layer. This is achieved by dividing the Blasius velocity profile by the corresponding momentum thickness Reynolds number and a constant of 2.193. In other words, the maximum of the profile is proportional to the momentum thickness Reynolds number and can therefore be related to the transition correlations [22] as follows:

$$Re_\theta = \frac{\max(Re_v)}{2.193}$$

(2)

Based on this observation, a general framework can be built, which can serve as a local environment for correlation based transition models.

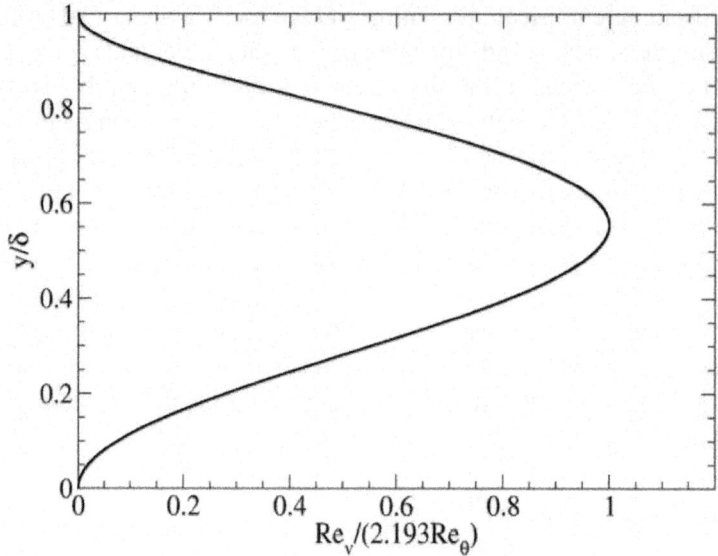

Figure. 1: Scaled vorticity Reynolds number (Re$_v$) profile in a Blasius boundary layer.

When the laminar boundary layer is subjected to strong pressure gradients, the relationship between momentum thickness and vorticity Reynolds number described by Equation (2) changes due to the change in the shape of the profile. The relative difference between momentum thickness and vorticity Reynolds number, as a function of shape factor (H), is shown in Figure 2. For moderate pressure gradients (2.3 < H < 2.9) the difference between the actual momentum thickness Reynolds number and the maximum of the vorticity Reynolds number is less than 10%. Based on boundary layer analysis a shape factor of 2.3 corresponds to a pressure gradient parameter (λ_θ) of approximately 0.06. Since the majority of experimental data on transition in favorable pressure gradients falls within that range (see for example reference [17]) the relative error between momentum thickness and vorticity Reynolds number is not of great concern under those conditions. For strong adverse pressure gradients the difference between the momentum thickness and vorticity Reynolds number can become significant, particularly near separation (H = 3.5). However, the trend with experiments is that adverse pressure gradients reduce the transition momentum thickness Reynolds number. In practice, if a constant transition momentum thickness Reynolds number is specified, the transition model is not very

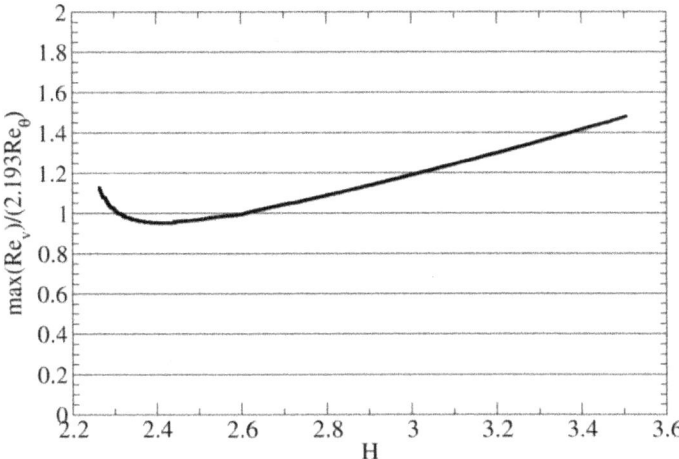

Figure. 2: Relative error between the maximum value of vorticity Reynolds number (Rev) and the momentum thickness Reynolds number (Reθ) as a function of boundary layer shape factor (H).

sensitive to adverse pressure gradients and an empirical correlation such as that of AbuGhannam and Shaw [17] is necessary in order to predict adverse pressure gradient transition accurately. In fact, the increase in vorticity Reynolds number with increasing shape factor can actually be used to predict separation induced transition. This is one of the main advantages of the present approach because the standard definition of momentum thickness Reynolds number is not suitable in separated flows. The function Re_v can be used on physical reasoning, by arguing that the combination of y2S is responsible for the growth of disturbances inside the boundary layer, whereas n = m r / is responsible for their damping. As y2S grows with the thickness of the boundary layer and m stays constant, transition will take place once a critical value of Rev is reached. The connection between the growth of disturbances and the function Re_v was shown by Van Driest and Blumer [21] in comparison with experimental data. As well, Langtry and Sjolander [15] found that the location in the boundary layer where Rev was largest corresponded surprisingly well to the location where the peak growth of disturbances was occurring, at least for bypass transition. The models proposed by Langtry & Sjolander [15] and Walters & Leylek, [16] use Re_v in physics-based arguments based on these observations of disturbance growth in the boundary layer during bypass transition. These models appear superior to conventional low-Re models, as they implicitly contain information of the thickness of the boundary layer. Nevertheless, the close integration of viscous sublayer damping and transition prediction does not easily allow for an independent calibration of both sub-models.

In the present approach first described in references [22, 29, 30 and 31] the main idea is to use a combination of the strain-rate Reynolds number with experimental transition correlations using standard transport equations. Due to the separation of viscous sublayer damping and transition prediction, the new method has provided the flexibility for introducing additional transition effects with relative ease. Currently, the main missing extensions are cross-flow instabilities and high-speed flow correlations and these do not pose any significant obstacles. The concept of linking the transition model with experimental data has proven to be an essential strength of the model and this is difficult to achieve with closures based on a physical modeling of these diverse phenomena. The present transition model is built on a transport equation for intermittency, which can be used to trigger transition locally. In addition to the transport equation for the intermittency, a second transport equation is solved for the transition onset momentum-thickness Reynolds number. This is required in order to capture the non-local influence of the turbulence intensity, which changes due to the decay of the turbulence kinetic energy in the free-stream, as well as due to changes in the free-stream velocity outside the boundary layer. This second transport equation is an essential part of the model as it ties the empirical correlation to the onset criteria in the intermittency equation. Therefore, it allows the model to be used in general geometries and over multiple airfoils, without additional information on the geometry. The intermittency function is coupled with the SST k-w based turbulence model [32]. It is used to turn on the production term of the turbulent kinetic energy downstream of the transition point based on the relation between transition momentumthickness and strain-rate Reynolds number. As the strain-rate Reynolds number is a local property, the present formulation avoids another very severe shortcoming of the correlation-based models, namely their limitation to 2D flows. It therefore allows the simulation of transition in 3D flows originating from different walls. The formulation of the intermittency has also been extended to account for the rapid onset of transition caused by separation of the laminar boundary layer (Equ. 17). In addition the model can be fully calibrated with internal or proprietary transition onset and transition length correlations. The correlations can also be extended to flows with rough walls or to flows with cross-flow instability. It should be stressed that the proposed transport equations do not attempt to model the physics of the transition process (unlike e.g. turbulence models), but form a framework for the implementation of correlation-based models into general-purpose CFD methods. In order to distinguish the present concept from physics based transition modeling, it is named LCTM – Local Correlation-based Transition Modeling.

Transition Model Equations

The present transition model formulation is described very briefly for completeness, a detailed description of the model and its development can be found in Langtry et al. [25]. It should be noted that a few changes have been made to the model since it was first published [29] in order to improve the predictions of natural transition. These include:

- A new transition onset correlation that results in improved predictions for both natural and bypass transition.
- A modification to the separation induced transition modification that prevents it from causing early transition near the separation point.
- Some adjustments of the model coefficients in order to better account for flow history effects on the transition onset location.

It was expected that different groups will make numerous improvements to the model and consequently a naming convention was introduced in reference [29] in order to keep track of the various model versions. The basic model framework (transport equations without any correlations) was called the g-Req transition model. The version number given in reference [29] was called CFX-v-1.0. Based on this naming convention, the present model with the above modifications will be referred to as the g-Req model, CFX-v-1.1. The present transition model is briefly summarized in the following pages. The transport equation for the intermittency, g, reads:

$$\frac{\partial(\rho\gamma)}{\partial t} + \frac{\partial(\rho U_j \gamma)}{\partial x_j} = P_\gamma - E_\gamma + \frac{\partial}{\partial x_j}\left[\left(\mu + \frac{\mu_t}{\sigma_f}\right)\frac{\partial\gamma}{\partial x_j}\right]$$

(3)

The transition sources are defined as follows:

$$P_{\gamma 1} = F_{length} c_{a1} \rho S \left[\gamma F_{onset}\right]^{0.5}\left(1 - c_{e1}\gamma\right)$$

(4)

where S is the strain rate magnitude. F_{length} is an empirical correlation that controls the length of the transition region. The destruction/relaminarization source is defined as follows:

$$E_\gamma = c_{a2}\rho\Omega\gamma F_{turb}\left(c_{e2}\gamma - 1\right)$$

(5)

where W is the vorticity magnitude. The transition onset is controlled by the following functions:

$$\mathrm{Re}_V = \frac{\rho y^2 S}{\mu}$$

(6)

$$F_{onset1} = \frac{Re_v}{2.193 \cdot Re_{\theta c}}$$

(7)

$$F_{onset2} = \min\left(\max\left(F_{onset1}, F_{onset1}{}^4\right), 2.0\right)$$

(8)

$$R_T = \frac{\rho k}{\mu \omega}$$

(9)

$$F_{onset3} = \max\left(1 - \left(\frac{R_T}{2.5}\right)^3, 0\right)$$

(10)

$$F_{onset} = \max\left(F_{onset2} - F_{onset3}, 0\right)$$

(11)

$Re_{\theta c}$ is the critical Reynolds number where the intermittency first starts to increase in the boundary layer. This occurs upstream of the transition Reynolds number, $Re_{\theta t}$ %, and the difference between the two must be obtained from an empirical correlation. Both the Flength and Reθc correlations are functions of $Re_{\theta t}$ %.

Based on the T3B, T3A, T3A- and the Schubauer and Klebanof test cases a correlation for F_{length} based on $Re_{\theta t}$ from an empirical correlation is defined as:

$$F_{length} = \begin{cases} \left[398.189 \cdot 10^{-1} + (-119.270 \cdot 10^{-4})\tilde{Re}_{\theta t} + (-132.567 \cdot 10^{-6})\tilde{Re}_{\theta t}{}^2\right], \tilde{Re}_{\theta t} < 400 \\ \left[263.404 + (-123.939 \cdot 10^{-2})\tilde{Re}_{\theta t} + (194.548 \cdot 10^{-5})\tilde{Re}_{\theta t}{}^2 + (-101.695 \cdot 10^{-8})\tilde{Re}_{\theta t}{}^3\right], 400 \le \tilde{Re}_{\theta t} < 596 \\ \left[0.5 - (\tilde{Re}_{\theta t} - 596.0) \cdot 3.0 \cdot 10^{-4}\right], 596 \le \tilde{Re}_{\theta t} < 1200 \\ \left[0.3188\right], 1200 \le \tilde{Re}_{\theta t} \end{cases}$$

(12)

In certain cases such as transition at higher Reynolds numbers the transport equation for $Re_{\theta t}$ % will often decrease to very small values in the boundary layer shortly after transition. Because F_{length} is based on $Re_{\theta t}$ % this can result in a local increase in the source term for the intermittency equation, which in turn can show up as a sharp increase in the skin friction. The skin friction does eventually return back to the fully turbulent value however this effect is unphysical. It appears to be caused by a sharp change in the y+ in the viscous sublayer where the intermittency decreases back to its minimum value due to the destruction term (Eq. 5). The effect can be eliminated by forcing F_{length} to always be equal to its maximum value (in this case 40.0) in the viscous sublayer. The modification for doing this is shown below. The modification does not appear to have any effect on the predicted transition length. An added benefit is that at higher Reynolds numbers the model now appears to predict the skin friction over shoot measured by experiments.

$$F_{sublayer} = e^{-\left(\frac{R_\omega}{0.4}\right)^2}$$

(13)

$$R_\omega = \frac{\rho y^2 \omega}{500\mu}$$

(14)

$$F_{length} = F_{length}\left(1 - F_{sublayer}\right) + 40.0 \cdot F_{sublayer}$$

(15)

The correlation between $Re_{\theta c}$ and $\tilde{Re}_{\theta t}$ % is defined as follows:

$$Re_{\theta c} = \begin{cases} \left[\tilde{Re}_{\theta t} - \left(396.035 \cdot 10^{-2} + (-120.656 \cdot 10^{-4})\tilde{Re}_{\theta t} + (868.230 \cdot 10^{-6})\tilde{Re}_{\theta t}^2 \right. \right. \\ \left. \left. +(-696.506 \cdot 10^{-9})\tilde{Re}_{\theta t}^3 + (174.105 \cdot 10^{-12})\tilde{Re}_{\theta t}^4\right)\right], \tilde{Re}_{\theta t} \leq 1870 \\ \left[\tilde{Re}_{\theta t} - \left(593.11 + \left(\tilde{Re}_{\theta t} - 1870.0\right) \cdot 0.482\right)\right], \tilde{Re}_{\theta t} > 1870 \end{cases}$$

(16)

The constants for the intermittency equation are:

$$c_{e1} = 1.0; \quad c_{a1} = 2.0; \qquad c_{e2} = 50; \quad c_{a2} = 0.06; \quad \sigma_f = 1.0;$$

The modification for separation-induced transition is:

$$\gamma_{sep} = \min\left(s_1 \max\left[0, \left(\frac{Re_v}{3.235 Re_{\theta c}}\right) - 1\right] F_{reattach}, 2\right) F_{\theta t}$$

(17)

$$F_{reattach} = e^{-\left(\frac{R_T}{20}\right)^4}$$

(18)

$$\gamma_{eff} = \max\left(\gamma, \gamma_{sep}\right)$$

(19)

$$s_1 = 2$$

(20)

The model constants in Equ. 17 have been adjusted from those of Menter et al. [31] in order to improve the predictions of separated flow transition. See Langtry [33] for a detailed discussion of the changes to the model from the Menter et al. [31] version. The main difference is the constant that controls the relation between Re_v and $Re_{\theta c}$ was changed from 2.193, it's value for a Blasius boundary layer, to 3.235, the value at a separation point where the shape factor (H) is 3.5 (see Figure 2). The boundary condition for g at a wall is zero normal flux while for an inlet g is equal to 1.0. An inlet g equal to 1.0 is necessary in order to preserve the original turbulence models freestream turbulence decay rate. The transport equation for the transition momentum thickness Reynolds number, $Re\theta t$ % , reads:

$$\frac{\partial\left(\rho\tilde{Re}_{\theta t}\right)}{\partial t} + \frac{\partial\left(\rho U_j \tilde{Re}_{\theta t}\right)}{\partial x_j} = P_{\theta t} + \frac{\partial}{\partial x_j}\left[\sigma_{\theta t}\left(\mu + \mu_t\right)\frac{\partial \tilde{Re}_{\theta t}}{\partial x_j}\right]$$

(21)

Outside the boundary layer, the source term $P_{\theta t}$ is designed to force the transported scalar $Re_{\theta t}$ % to match the local value of $Re_{\theta t}$ calculated from the empirical correlation (Equ. 35, 36). The source term is defined as follows:

$$P_{\theta t} = c_{\theta t}\frac{\rho}{t}\left(Re_{\theta t} - \tilde{Re}_{\theta t}\right)\left(1.0 - F_{\theta t}\right)$$

(22)

$$t = \frac{500\mu}{\rho U^2}$$

(23)

where t is a time scale, which is present for dimensional reasons. The time scale was determined based on dimensional analysis with the main criteria being that it had to scale with the convective and diffusive terms in the transport equation. The blending function $F_{\theta t}$ is used to turn off the source term in the boundary layer and allow the transported scalar $Re_{\theta t}$ % to diffuse in from the freestream. $F_{\theta t}$ is equal to zero in the freestream and one in the boundary layer. The $F_{\theta t}$ blending function is defined as follows:

$$F_{\theta t} = min\left(max\left(F_{wake}\cdot e^{-\left(\frac{y}{\delta}\right)^4}, 1.0 - \left(\frac{\gamma - 1/c_{e2}}{1.0 - 1/c_{e2}}\right)^2\right), 1.0\right)$$

(24)

$$\theta_{BL} = \frac{\tilde{Re}_{\theta t}\mu}{\rho U}; \quad \delta_{BL} = \frac{15}{2}\theta_{BL}; \quad \delta = \frac{50\Omega y}{U}\cdot\delta_{BL}$$

(25)

$$Re_\omega = \frac{\rho\omega y^2}{\mu}; \quad F_{wake} = e^{-\left(\frac{Re_\omega}{1E+5}\right)^2}$$

(26)

The F_{wake} function ensures that the blending function is not active in the wake regions downstream of an airfoil/blade.

The model constants for the $\tilde{Re}_{\theta t}$ % equation are:

$$c_{\theta t} = 0.03; \quad \sigma_{\theta t} = 2.0$$

(27)

The boundary condition for $\tilde{Re}_{\theta t}$ % at a wall is zero flux. The boundary condition for $\tilde{Re}_{\theta t}$ % at an inlet should be calculated from the empirical correlation (Equ. 35, 36) based on the inlet turbulence intensity. The empirical correlation for transition onset is based on the following parameters:

$$\lambda_\theta = \frac{\rho\theta^2}{\mu}\frac{dU}{ds} \tag{28}$$

$$Tu = 100\frac{\sqrt{2k/3}}{U} \tag{29}$$

Where dU/ds is the acceleration along the streamwise direction and can be computed by taking the derivative of the velocity (U) in the x, y and z directions and then summing the contribution of these derivatives along the streamwise flow direction:

$$U = \left(u^2 + v^2 + w^2\right)^{\frac{1}{2}} \tag{30}$$

$$\frac{dU}{dx} = \frac{1}{2}\left(u^2 + v^2 + w^2\right)^{-\frac{1}{2}} \cdot \left[2u\frac{du}{dx} + 2v\frac{dv}{dx} + 2w\frac{dw}{dx}\right] \tag{31}$$

$$\frac{dU}{dy} = \frac{1}{2}\left(u^2 + v^2 + w^2\right)^{-\frac{1}{2}} \cdot \left[2u\frac{du}{dy} + 2v\frac{dv}{dy} + 2w\frac{dw}{dy}\right] \tag{32}$$

$$\frac{dU}{dz} = \frac{1}{2}\left(u^2 + v^2 + w^2\right)^{-\frac{1}{2}} \cdot \left[2u\frac{du}{dz} + 2v\frac{dv}{dz} + 2w\frac{dw}{dz}\right] \tag{33}$$

$$\frac{dU}{ds} = \left[(u/U)\frac{dU}{dx} + (v/U)\frac{dU}{dy} + (w/U)\frac{dU}{dz}\right] \tag{34}$$

The use of the streamline direction is not Galilean invariant. However, this deficiency is inherent to all correlation-based models, as their main variable, the turbulence intensity is already based on the local freestream velocity and does therefore violate Galilean invariance. This is not problematic, as the correlations are defined with respect to a wall boundary layer and all velocities are therefore relative to the wall. Nevertheless, multiple moving walls in one domain will likely require additional information.

The use of the streamline direction is not Galilean invariant. However, this deficiency is inherent to all correlation-based models, as their main variable, the turbulence intensity is already based on the local freestream velocity and does therefore violate Galilean invariance. This is not problematic, as the correlations are defined with respect to a wall boundary layer and all velocities are therefore relative to the wall. Nevertheless, multiple moving walls in one domain will likely require additional information. The empirical correlation has been modified from reference [29] to improve the predictions of natural transition. The predicted transition Reynolds number as a function of turbulence intensity is shown in Figure 3. For pressure gradient flows the model predictions are similar to the Abu-Ghannam and Shaw [17] correlation. The empirical correlation is defined as follows:

$$Re_{\theta t} = \left[1173.51 - 589.428 Tu + \frac{0.2196}{Tu^2} \right] F(\lambda_\theta), Tu \leq 1.3 \tag{35}$$

$$Re_{\theta t} = 331.50 [Tu - 0.5658]^{-0.671} F(\lambda_\theta), Tu > 1.3 \tag{36}$$

$$F(\lambda_\theta) = 1 - \left[-12.986\lambda_\theta - 123.66\lambda_\theta{}^2 - 405.689\lambda_\theta{}^3 \right] e^{-\left[\frac{Tu}{1.5} \right]^{1.5}}, \lambda_\theta \leq 0 \tag{37}$$

$$F(\lambda_\theta) = 1 + 0.275 \left[1 - e^{[-35.0\lambda_\theta]} \right] e^{\left[\frac{-Tu}{0.5} \right]}, \lambda_\theta > 0 \tag{38}$$

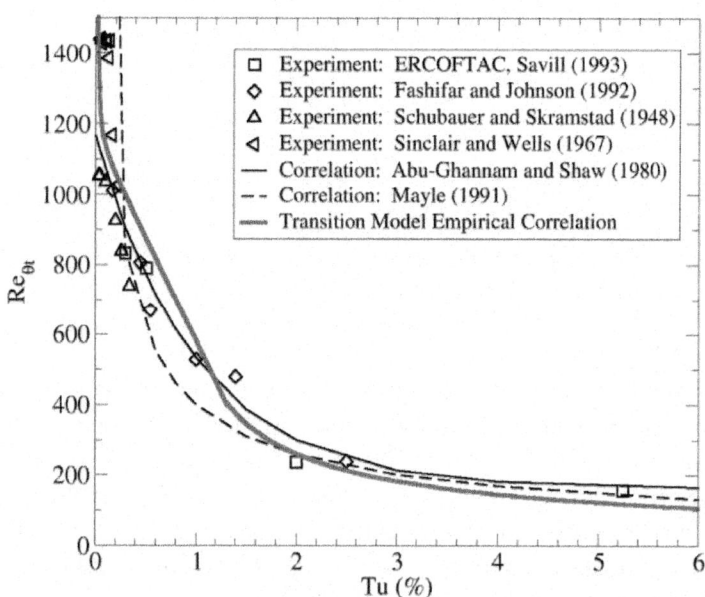

Figure. 3: Transition onset momentum thickness Reynolds number (Re$_{\theta t}$) predicted by the new correlation as a function of turbulence intensity (Tu) for a flat plate with zero pressure gradient.

For numerical robustness the acceleration parameters, the turbulence intensity and the empirical correlation should be limited as follows:

$-0.1 \le \lambda_\theta \le 0.1$

$Tu \ge 0.027$

$\mathrm{Re}_{\theta t} \ge 20$

A minimum turbulence intensity of 0.027 percent results in a transition momentum thickness Reynolds number of 1450, which is the largest experimentally observed flat plate transition Reynolds number based on the Sinclair and Wells [36] data. For cases where larger transition Reynolds are believed to occur (e.g. aircraft in flight) this limiter may need to be adjusted downwards. The empirical correlation is used only in the source term (Eq. 22) of the transport equation for the transition onset momentum thickness Reynolds number. Equations 35 to 38 must be solved iteratively because the momentum thickness (θ_t) is present in the left hand side of the equation and also in the right hand side in the pressure gradient parameter (λ_q). In the present work an initial guess for the local value of θ_t was obtained based on the zero pressure gradient solution of Eq. 35, 36 and the local values of U, r and m. With this initial guess, equations 35 to 38 were solved by iterating on the value of θ_t and convergence was obtained in less then ten iterations using a shooting point method. The transition model interacts with the SST turbulence model [32], as follows:

$$\frac{\partial}{\partial t}(\rho k) + \frac{\partial}{\partial x_j}(\rho u_j k) = \tilde{P}_k - \tilde{D}_k + \frac{\partial}{\partial x_j}\left[(\mu + \sigma_k \mu_t)\frac{\partial k}{\partial x_j}\right]$$

(39)

$$\tilde{P}_k = \gamma_{eff} P_k; \quad \tilde{D}_k = \min\left(\max(\gamma_{eff}, 0.1), 1.0\right) D_k$$

(40)

$$R_y = \frac{\rho y \sqrt{k}}{\mu}; \quad F_3 = e^{-\left(\frac{R_y}{120}\right)^8}; \quad F_1 = \max\left(F_{1orig}, F_3\right)$$

(41)

where P_k and D_k are the original production and destruction terms for the SST model and F1orig is the original SST blending function. Note that the production term in the w-equation is not modified. The rationale behind the above model formulation is given in detail in reference [29]. In order to capture the laminar and transitional boundary layers correctly, the grid must have a y+ of approximately one at the first grid point off the wall. If the y^+ is too large (i.e. > 5) then the transition onset location moves upstream with increasing y^+. All simulations have been performed with CFX-5 using a bounded second order upwind biased discretisation for the mean flow, turbulence and transition equations.

TEST CASES

The remaining part of the chapter will give an overview of some of the public-domain testcases which have been computed with the model described above. This naturally requires a compact representation of the simulations. Most of the cases are described in far more detail in reference [33], including grid refinement and sensitivity studies.

Flat Plate Test Cases

The flat plate test cases that where used to calibrate the model are the ERCOFTAC T3 series of flat plate experiments [12, 13] and the Schubauer and Klebanof [37] flat plate experiment, all of which are commonly used as benchmarks for transition models. Also included is a test case where the boundary layer experiences a strong favorable pressure gradient that causes it to relaminarize [38]. The inlet conditions for these testcases are summarized in Table 1. The three cases T3A-, T3A, and T3B have zero pressure gradients with different freestream turbulence intensity (FSTI) levels corresponding to transition in the bypass regime.

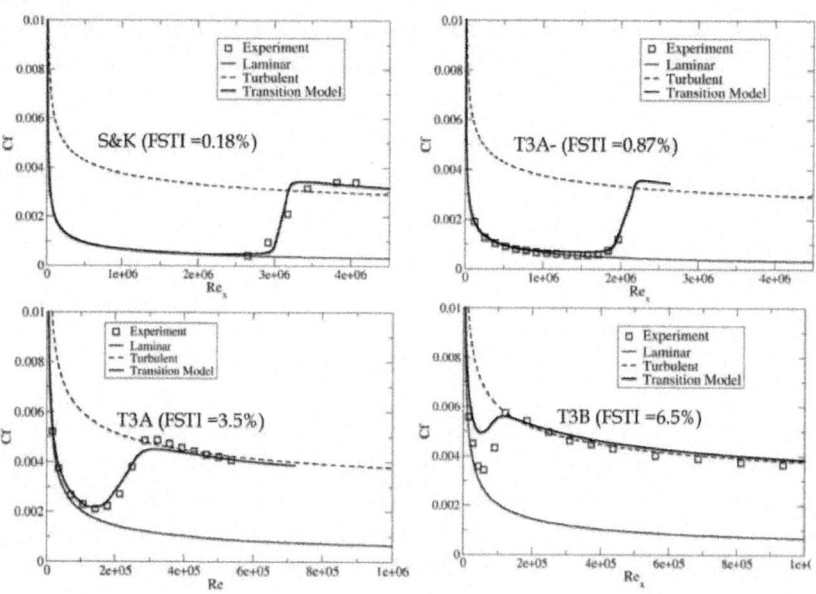

Figure. 4: Results for flat plate test cases with different freestream turbulence levels (FSTI – Freestream Turbulence Intensity).

The Schubauer and Klebanof (S&K) test case has a low free-stream turbulence intensity and corresponds to natural transition. Figure 4 shows the comparison of the model prediction with experimental data for theses cases. It also gives the corresponding FSTI values. In all simulations, the inlet turbulence levels were specified to match the experimental turbulence intensity and its decay rate. This was done by fixing the inlet turbulence intensity and via trial and error adjusting the inlet viscosity ratio (i.e. the w inlet condition) to match the experimentally measured turbulence levels at various downstream locations. As the freestream turbulence increases, the transition location moves to lower Reynolds numbers.

Table 1: Inlet condition for the flat plate test cases

Case	Inlet Velocity (m/s)	Turbulence Intensity (%) Inlet / Leading Edge value	μ_t/μ	Density (kg/m³)	Dynamic Viscosity (kg/ms)
T3A	5.4	3.3	12.0	1.2	1.8×10^{-5}
T3B	9.4	6.5	100.0	1.2	1.8×10^{-5}
T3A-	19.8	0.874	8.72	1.2	1.8×10^{-5}
Schubauer and Klebanof	50.1	0.3	1.0	1.2	1.8×10^{-5}
T3C2	5.29	3.0	11.0	1.2	1.8×10^{-5}
T3C3	4.0	3.0	6.0	1.2	1.8×10^{-5}
T3C4	1.37	3.0	8.0	1.2	1.8×10^{-5}
T3C5	9.0	4.0	15.0	1.2	1.8×10^{-5}
Relaminarization	1.4	5.5	15	1.2	1.8×10^{-5}

The T3C test cases consist of a flat plate with a favorable and adverse pressure gradient imposed by the opposite converging/diverging wall. The wind tunnel Reynolds number was varied for the four cases (T3C5, T3C3, T3C2, T3C4) thus moving the transition location from the favorable pressure at the beginning of the plate to the adverse pressure gradient at the end. The cases are used to demonstrate the transition models ability to predict transition under the influence of various pressure gradients. Figure 5 details the results for the

pressure gradient cases. The effect of the pressure gradient on the transition length is clearly visible with favorable pressure gradients increasing the transition length and adverse pressure gradients reducing it. For the T3C4 case the laminar boundary layer actually separates and undergoes separation induced transition. The relaminarization test case is shown in Figure 6. For this case the opposite converging wall imposes a strong favorable pressure gradient that can relaminarize a turbulent boundary layer. In both the experiment and in the CFD prediction the boundary layer was tripped near the plate leading edge. In the CFD computation this was accomplished by injecting a small amount of turbulent air into the boundary layer with a turbulence intensity of 3%. The same effect could have been accomplished with a small step or gap in the CFD geometry. Downstream of the trip the boundary layer slowly relaminarizes due to the strong favorable pressure gradient. For all of the flat plate test cases the agreement with the data is generally good, considering the diverse nature of the physical phenomena computed, ranging from bypass transition to natural transition, separation-induced transition and even relaminarization.

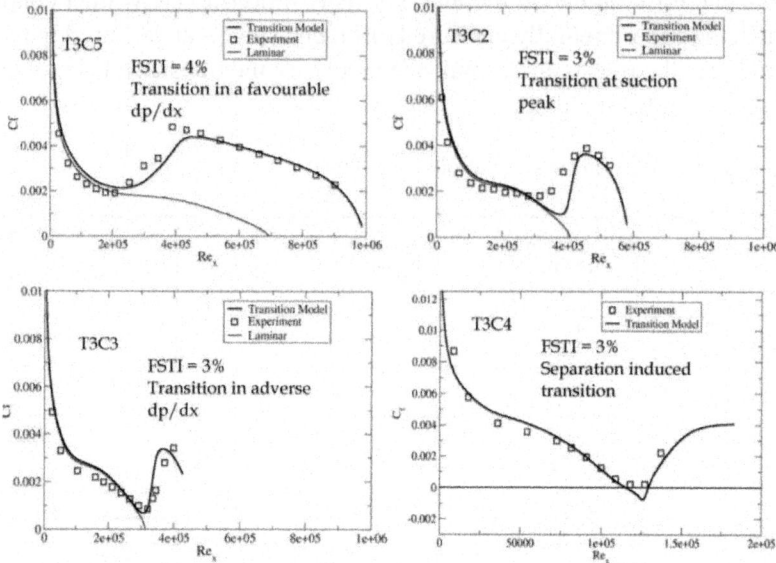

Figure. 5: Results for flat plate test cases where variation of the tunnel Reynolds number causes transition to occur in different pressure gradients (dp/dx).

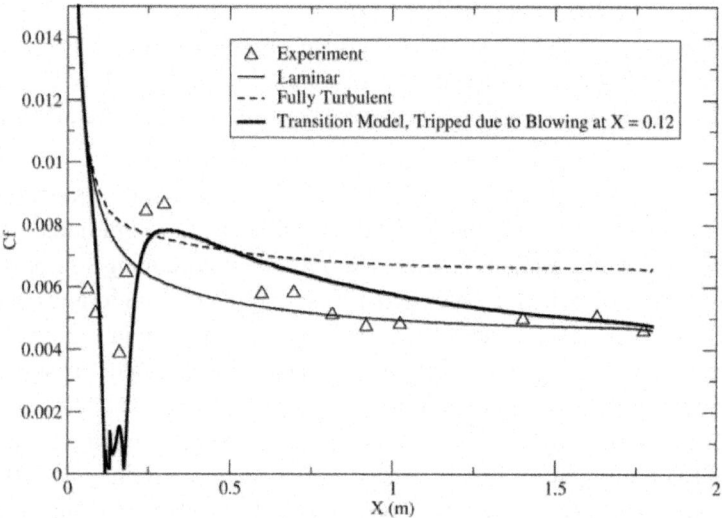

Figure. 6: Predicted skin friction (Cf) for a flat plate with a strong acceleration that causes the boundary layer to relaminarize.

Turbomachinery Test Cases

This section descrives a few of the turbomachinery test cases that have been used to validate the transition model including a compressor blade, a low-pressure turbine and a high pressure turbine. A summary of the inlet conditions is shown in Table 2. For the Zierke and Deutsch [39] compressor blade, transition on the suction side occurs at the leading edge due to a small leading edge separation bubble on the suction side. On the pressure side, transition occurs at about mid-chord. The turbulence contours and the skin

Table 2: Inlet conditions for the turbomachinery test cases

Case	$Re_x = \rho c U_0/\mu$ $(\times 10^6)$	Mach $= U_0/a$ where speed of sound $(a) = (\gamma RT)^{0.5}$	Chord (c) (m)	FSTI (%)	μ_t/μ
Zierke and Deutsch Compressor Incidence = -1.5°	0.47	0.1	0.2152	0.18	2.0
Pak-B Low-Pressure Turbine Blade	0.05, 0.075, 0.1	0.03	0.075	0.08, 2.35, 6.0	6.5 - 30
VKI MUR Transonic Guide Vane	0.26	Inlet: 0.15 Outlet: 1.06	0.037	1.0, 6.5	11, 1000

friction distribution are shown in Figure 7. There appears to be a significant amount of scatter in the experimental data; however, overall the transition model is predicting the major flow features correctly (i.e. fully turbulent suction side, transition at mid-chord on the pressure side). One important issue to note is the effect of stream-wise grid resolution on resolving the leading edge laminar separation and subsequent transition on the suction side. If the number of stream-wise nodes clustered around the leading edge is too low, the model cannot resolve the rapid transition and a laminar boundary layer on the suction side is the result. For the present study, 60 streamwise nodes were used between the leading edge and the x/C = 0.1 location. The Pratt and Whitney PAK-B low pressure turbine blade is a particularly interesting airfoil because it has a loading profile similar to the rotors found in many modern aircraft engines [40]. The low-pressure rotors on modern aircraft engines are extremely challenging flow fields. This is because in many cases the transition occurs in the free shear layer of a separation bubble on the suction side [4]. The onset of transition in the free shear layer determines whether or not the separation bubble will reattach as a turbulent boundary layer and, ultimately, whether or not the blade will stall. The present transition model would therefore be of great interest to turbine designers if it can accurately predict the transition onset location for these types of flows. Huang et al. [41] conducted experiments on the PAK-B blade cascade for a range of Reynolds numbers and turbulence intensities. The experiments were performed at the design incidence angle for Reynolds numbers of 50,000, 75,000, and 100,000 based on inlet velocity and axial chord length, with turbulence intensities of 0.08%, 2.35% and 6.0% (which corresponded to values of 0.08%, 1.6%, and 2.85% at the leading edge of the blade). The computed pressure coefficient distributions obtained with the transition model and fully turbulent model are compared to the experimental data for the 75 000 Reynolds number, 2.35% turbulent intensity case in Figure 8. On the suction side, a pressure plateau due to a laminar separation with turbulent reattachment exists. The fully turbulent computation completely misses this phenomenon because the boundary layer remains attached over the entire length of the suction surface. The transition model can predict the pressure plateau due to the laminar separation and the subsequent turbulent reattachment location. The pressure side was predicted to be fully attached and laminar.

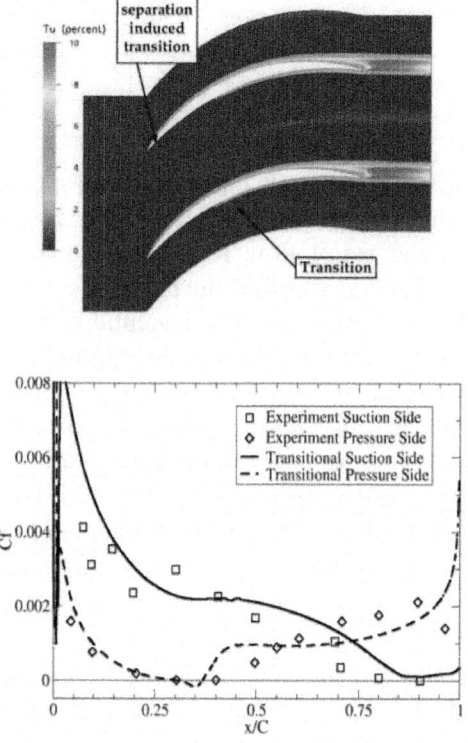

Figure. 7: Turbulence intensity contours (top) and cf-distribution against experimental data (right) for the Zierke & Deutsch compressor.

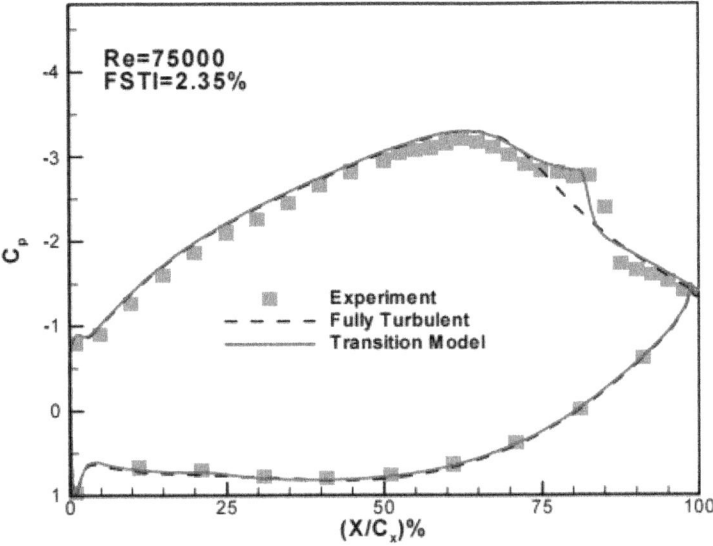

Figure. 8: Predicted blade loading for the Pak-B Low-Pressure turbine at a Reynolds number of 75 000 and a freestream turbulent intensity (FSTI) of 2.35%.

The computed pressure coefficient distributions for various Reynolds numbers and freestream turbulence intensities compared to experimental data are shown in Figure 9. In this figure, the comparisons are organized such that the horizontal axis denotes the Reynolds number whereas the vertical axis corresponds to the freestream turbulence intensity of the specific case. As previously pointed out, the most important feature of this test case is the extent of the separation bubble on the suction side, characterized by the plateau in the pressure distribution. The size of the separation bubble is actually a complex function of the Reynolds number and the freestream turbulence value. As the Reynolds number or freestream turbulence decrease, the size of the separation and hence the pressure plateau increases. The computations with the transition model compare well with the experimental data for all of the cases considered, illustrating the ability of the model to capture the effects of Reynolds number and turbulence intensity variations on the size of a laminar separation bubble and the subsequent turbulent reattachment. The surface heat transfer for the transonic VKI MUR 241 (FSTI = 6.0%) and MUR 116 (FSTI = 1.0%) test cases [42] is shown in Figure 10. The strong acceleration on the suction side for the MUR 241 case keeps the flow laminar until a weak shock at mid chord, whereas for the MUR 116 case the flow is laminar until right before the trailing edge. Downstream of transition there appears to be a significant difference between the predicted turbulent heat transfer and the measured value. It is possible that this is the result of a Mach number (inlet Mach number M_{ainlet} =0.15, Ma_{outlet} =1.089) effect on the transition length [43]. At present, no attempt has been made to account for this effect in the model. It can be incorporated in future correlations, if found consistently important.

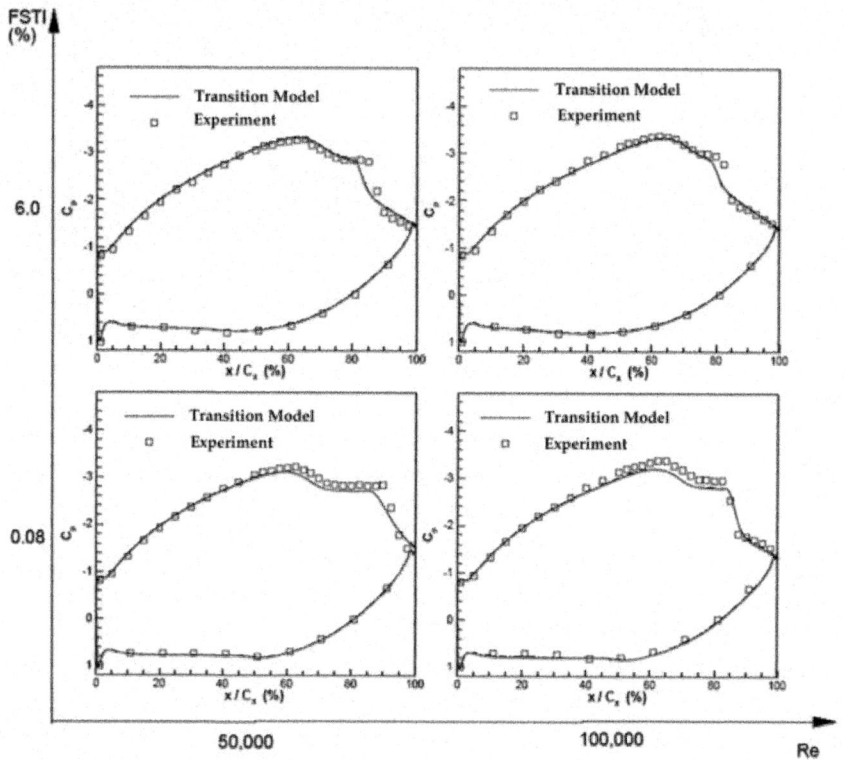

Figure. 9: Blade loading for the Pak-B Low-Pressure turbine at various freestream turbulence intensities (FSTI) and Reynolds numbers (Re).

The pressure side heat transfer is of particular interest for this case. For both cases, transition did not occur on the pressure side, however, the heat transfer was significantly increased for the high turbulence intensity case. This is a result of the large freestream levels of turbulence which diffuse into the laminar boundary layer and increase the heat transfer and skin friction. From a modeling standpoint, the effect was caused by the large freestream viscosity ratio necessary for MUR 241 to keep the turbulence intensity from decaying below 6%, which is the freestream value quoted in the experiment. The enhanced heat transfer on the pressure side was also present in the experiment and the effect appears to be physical. The model can predict this effect, as the intermittency does not multiply the eddy-viscosity but only the production term of the k-equation. The diffusive terms are therefore active in the laminar region. The S809 airfoil is a 21% thick, laminar-flow airfoil that was designed specifically for horizontal-axis wind turbine (HAWT) applications. The airfoil profile is shown in Figure 11. The experimental results where obtained in the low-turbulence wind tunnel at the Delft

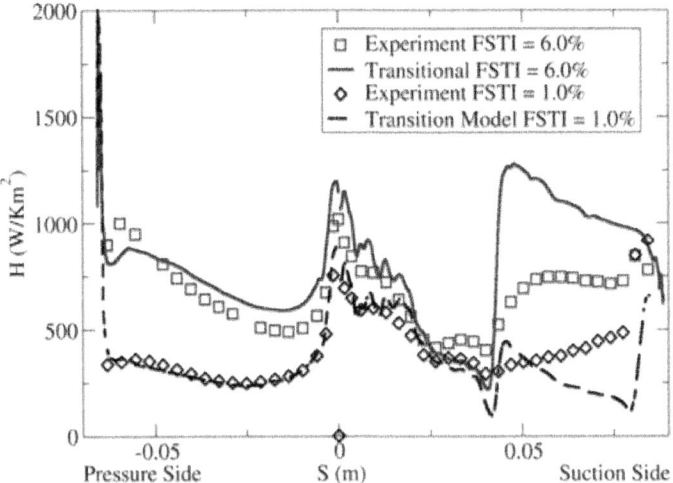

Figure. 10: Heat transfer for the VKI MUR241 (FSTI = 6.0%) and MUR116 (FSTI = 1.0%) test cases.

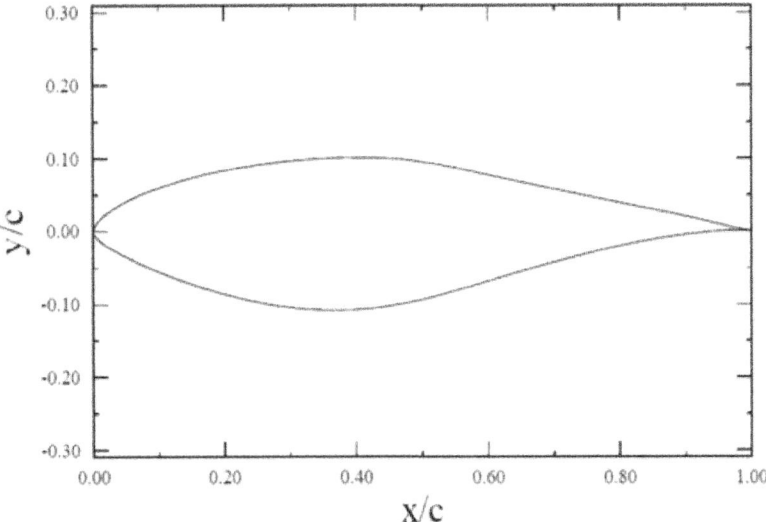

Figure. 11: S809 Airfoil Profile.

University of Technology [44, 45]. The detailed CFD results can be found in reference [46]. The predicted pressure distribution around the airfoil for angles of attack (AoA) of 1° is shown in Figure 12. For the 1° AoA case the flow is laminar for the first 0.5 chord of the airfoil on both the suction and pressure sides. The boundary layers then undergo a laminar separation and reattach as a turbulent boundary layer and this is clearly visible in the experimental pressure distribution plateaus. The fully turbulent computation obviously

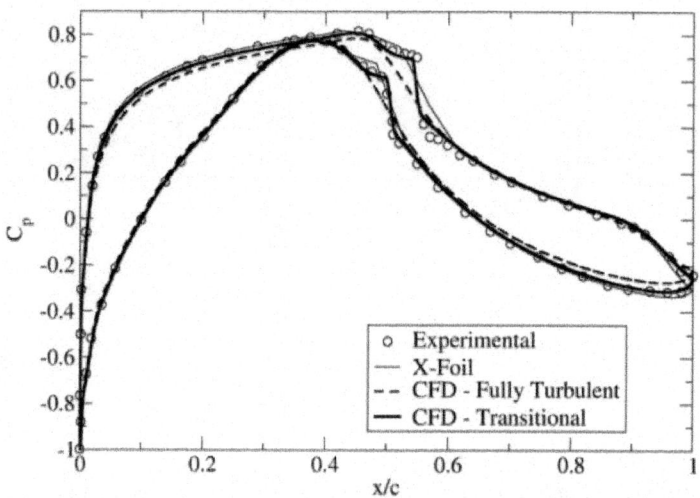

Figure. 12: Pressure distribution (Cp) for the S809 airfoil at 1° angle of attack.

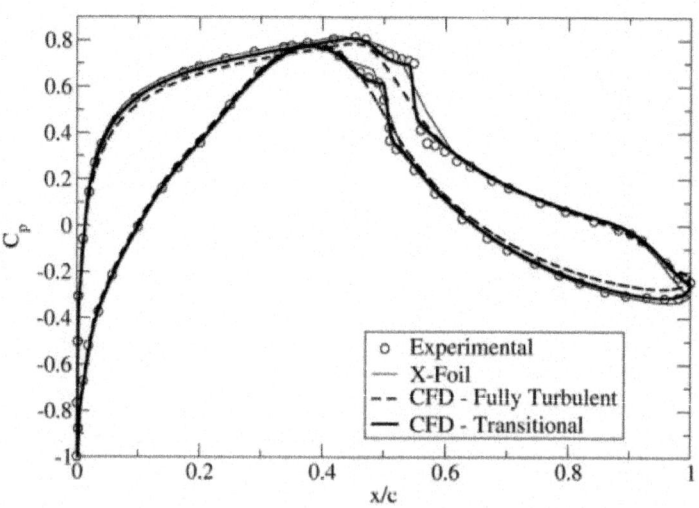

Figure. 12: Pressure distribution (Cp) for the S809 airfoil at 1° angle of attack.

does not capture this phenomenon, as the turbulent boundary layers remain completely attached. Both the transitional CFD and X-Foil solutions do predict the laminar separation bubble. However, X-Foil appears to slightly over predict the reattachment location while the transitional CFD simulation is in very good agreement with the experiment.

Table 3: Inlet conditions for the S809 test case

Case	Re_x (x10^6)	Mach	Chord (m)	FSTI (%)	μ_t/μ
S809 Airfoil	2.0	0.1	1	0.2	10

The predicted transition locations as a function of angle of attack are shown in Figure 13. The experimental transition locations were obtained using a stethoscope method (Somers, [42]). In general the present transition model would appear to be in somewhat better agreement with the experiment than the X-Foil code, particularly around 14° angle of attack. However, at the moderate angles of attack all of the results appear be to within approximately 5% chord of each other. The X-Foil transition locations appear to change quite rapidly over a few degrees angle of attack while the transition model has a much smoother change in the transition location. The experimental data would appear to confirm that the smooth change in transition location is more physical, however this observation is based primarily on the 10° and 14° angle of attack cases. The results obtained for the lift and drag polars are shown in Figures 14 and 15. Between 0° and 9° the lift coefficients (Cl) predicted by the transitional CFD results are in very good agreement with the experiment while both the XFoil and fully turbulent CFD and results appear to under-predict the lift curve by approximately 0.1. As well, between 0° and 9° the drag coefficient (Cd) predicted by the transitional CFD and X-Foil results are in very good agreement with the experiment while the fully turbulent CFD simulation significantly over predicts the drag, as expected.

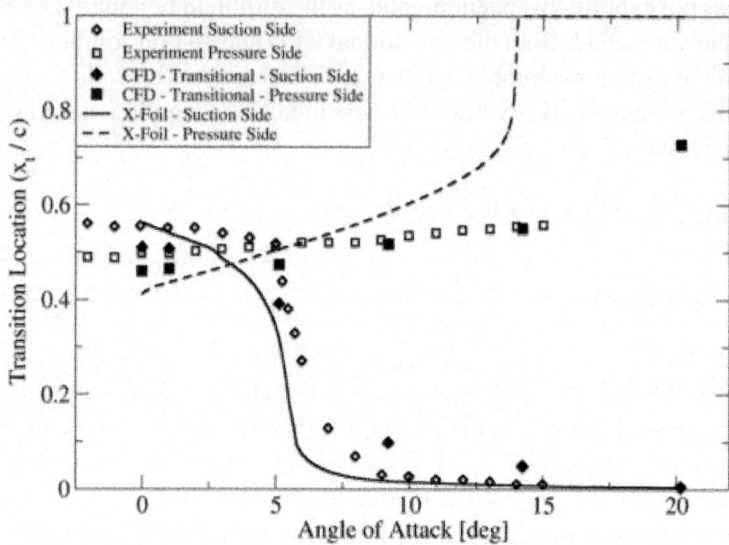

Figure. 13: Transition location (xt/c) vs angle of attack for the S809 airfoil.

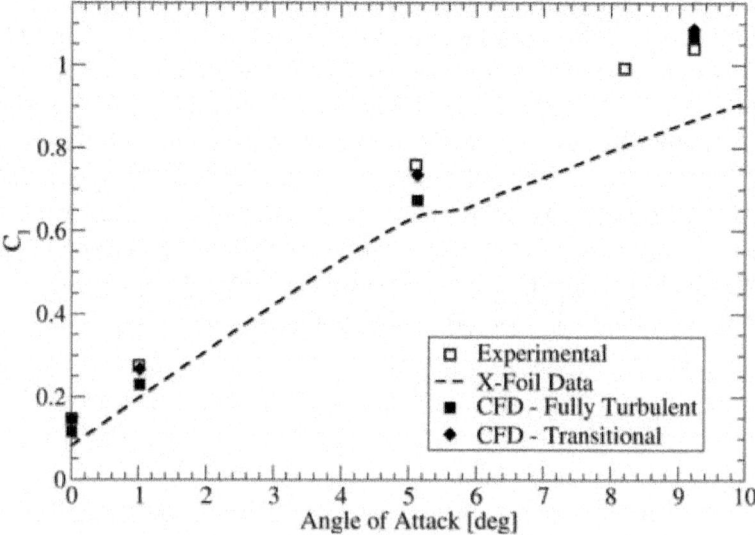

Figure. 14: Lift Coefficient (Cl) Polar for the S809 airfoil.

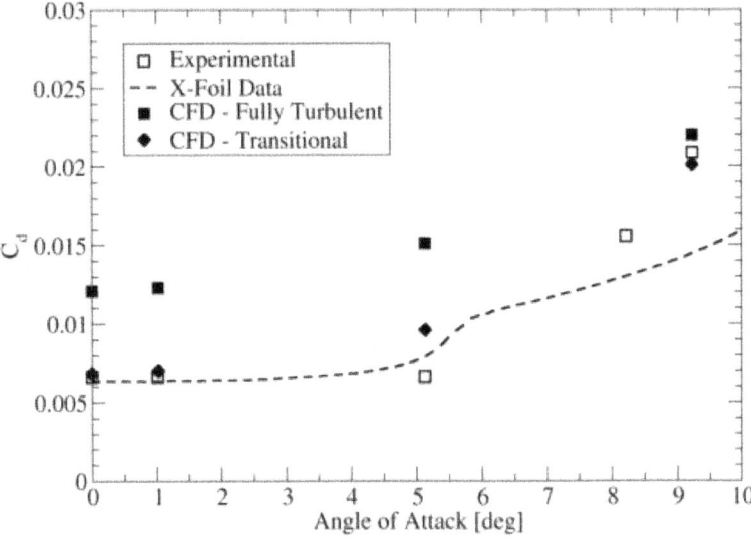

Figure. 15: Drag Coefficient (Cd) Polar for the S809 airfoil The results obtained for the lift and drag polars are.

CONCLUSIONS

In this chapter various methods for transition prediction in general purpose CFD codes have been discussed. In addition, the requirements that a model has to satisfy to be suitable for implementation into a general purpose CFD code have been listed. The main criterion is that non-local operations must be avoided. A new concept of transition modeling termed Local Correlation-based Transition Model (LCTM) was introduced. It combines the advantages of locally formulated transport equations with the physical information contained in empirical correlations. The g-Re$_{q\,t}$ransition model is a representative of that modeling concept. The model is based on two new transport equations (in addition to the k and w equations), one for intermittency and one for a transition onset criterion in terms of momentum thickness Reynolds number. The proposed transport equations do not attempt to model the physics of the transition process (unlike e.g. turbulence models), but form a framework for the implementation of transition correlations into general-purpose CFD methods. An overview of the g-Re$_q$ model formulation has been given along with the publication of the full model including some previously undisclosed empirical correlations that control the predicted transition length. The main goal of the present chapter was to publish the full model and release it to the research community so that it can continue to be further validated and possibly extended. Included in this chapter are a number of test cases that can be used to validate the implementation of the model in a given CFD code. The present transition model accounts for transition due to freestream turbulence intensity, pressure gradients and separation. It is fully CFD-compatible and does not negatively affect the convergence of the solver. Current limitations of the model are that crossflow instability or roughness are not included in the correlations and that the transition correlations are formulated non-Galilean invariant. These limitations are currently being investigated and can be removed in principle An overview of the test cases computed with the new model has been given. Due to the nature of the chapter, the presentation of each individual test case had to be brief. More details on the test case set-up, boundary conditions grid resolutions etc. can be found in the references. The purpose of the overview was to show that the model can handle a wide variety of geometries and physically diverse problems. The authors believe that the current model is a significant step forward in engineering transition modeling. Through the use of transport equations instead of search or line integration algorithms, the model formulation offers a flexible environment for engineering transition predictions that is fully compatible with the infrastructure of modern CFD methods. As a result, the model can be used in any general purpose CFD method without special provisions for geometry and grid topology. The authors believe that the LCTM concept of combining transition correlations with locally formulated transport equations has a strong potential for allowing the 1st order effects of transition to be included into today's industrial CFD simulations.

ACKNOWLEDGMENTS

The model development and validation at ANSYS CFX was funded by GE Aircraft Engines and GE Global Research. The authors would like to thank Dr. Stefan Voelker and Dr. Bill Solomon of GE for their support and numerous thoughtful discussions throughout the course of the model development. As well, Prof. G. Huang of Wright State University and Prof. B. Suzen of North Dakota State University who have supported the original model development with their extensive know-how and their in-house codes. Finally, the authors would also like to thank Dr. Chris Rumsey from the NASA Langley Research Center for supplying the geometry and experimental data for the McDonald Douglas 30P-30N flap and Dr. Helmut Sobieczky of the DLR for his helpful discussions on the DLR F-5 testcase.

REFERENCES

1.   Schlichting, H., 1979, Boundary Layer Theory, McGraw-Hill, 7th edition.

2.   Morkovin, M.V., 1969, "On the Many Faces of Transition", Viscous Drag Reduction, C.S. Wells, ed., Plenum Press, New York, pp 1-31.

3. Malkiel, E. and Mayle, R.E., 1996, "Transition in a Separation Bubble," ASME Journal of Turbo machinery, Vol. 118, pp. 752-759.

4. Mayle, R.E., 1991, "The Role of Laminar-Turbulent Transition in Gas Turbine Engines," Journal of Turbomachinery, Vol. 113, pp. 509-537.

5. Smith, A.M.O. and Gamberoni, N., 1956, "Transition, Pressure Gradient and Stability Theory," Douglas Aircraft Company, Long Beach, Calif. Rep. ES 26388.

6. Van Ingen, J.L., 1956, "A suggested Semi-Empirical Method for the Calculation of the Boundary Layer Transition Region," Univ. of Delft, Dept. Aerospace Engineering, Delft, The Netherlands, Rep. VTH-74. www.intechopen.com 56 Low Reynolds Number Aerodynamics and Transition

7. Stock, H.W. and Haase, W., 2000, "Navier-Stokes Airfoil Computations with eN Transition Prediction Including Transitional Flow Regions," AIAA Journal, Vol. 38, No. 11, pp. 2059 – 2066.

8. Drela, M., and Giles, M. B., 1987, "Viscous-Inviscid Analysis of Transonic and Low Reynolds Number Airfoils", AIAA Journal, Vol. 25, pp. 1347 – 1355.

9. Youngren, H., and Drela, M., 1991, "Viscous-Inviscid Method for Preliminary Design of Transonic Cascades", AIAA Paper No. 91-2364.

10. Jones, W. P., and Launder, B. E., 1973, "The Calculation of Low Reynolds Number Phenomena with a Two-Equation Model of Turbulence", Int. J. Heat Mass Transfer, Vol. 15, pp. 301-314.

11. Rodi, W. and Scheuerer, G., 1984, "Calculation of Laminar-Turbulent Boundary Layer Transition on Turbine Blades", AGARD CP 390 on Heat transfer and colloing in gas turbines.

12. Savill, A.M., 1993, Some recent progress in the turbulence modeling of by-pass transition, In: R.M.C. So, C.G. Speziale and B.E. Launder, Eds.: Near-Wall Turbulent Flows, Elsevier, p. 829.

13. Savill, A.M., 1996, One-point closures applied to transition, Turbulence and Transition Modeling, M. Hallbäck et al., eds., Kluwer, pp. 233-268.

14. Wilcox, D.C.W. (1994). Simulation of transition with a two-equation turbulence model, AIAA J. Vol. 32, No. 2.

15. Langtry, R.B., and Sjolander, S.A., 2002, "Prediction of Transition for Attached and Separated Shear Layers in Turbomachinery", AIAA Paper 2002-3643.

16. Walters, D.K and Leylek, J.H., 2002, "A New Model for Boundary-Layer Transition Using a Single-Point RANS Approach", ASME IMECE'02, IMECE2002-HT-32740.

17. Abu-Ghannam, B.J. and Shaw, R., 1980, "Natural Transition of Boundary Layers -The Effects of Turbulence, Pressure Gradient, and Flow History," Journal of Mechanical Engineering Science, Vol. 22, No. 5, pp. 213 – 228.

18. Mayle, R.E., 1991, "The Role of Laminar-Turbulent Transition in Gas Turbine Engines," Journal of Turbomachinery, Vol. 113, pp. 509-537.

19. Suzen, Y.B., Huang, P.G., Hultgren, L.S., Ashpis, D.E., 2003, "Predictions of Separated and Transitional Boundary Layers Under Low-Pressure Turbine Airfoil Conditions Using an Intermittency Transport Equation," Journal of Turbomachinery, Vol. 125, No. 3, July 2003, pp. 455-464.

20. Durbin, P.A., Jacobs, R.G. and Wu, X., 2002, "DNS of Bypass Transition," Closure Strategies for Turbulent and Transitional Flows, edited by B.E. Launder and N.D. Sandham, Cambridge University Press, pp. 449-463.

21. Van Driest, E.R. and Blumer, C.B., 1963, "Boundary Layer Transition: Freestream Turbulence and Pressure Gradient Effects," AIAA Journal, Vol. 1, No. 6, June 1963, pp. 1303-1306.

22. Menter, F.R., Esch, T. and Kubacki, S., 2002, "Transition Modelling Based on Local Variables", 5th International Symposium on Engineering Turbulence Modelling and Measurements, Mallorca, Spain.

23. Walters K. Cokljat, D., (2008), "A Three-Equation Eddy-Viscosity Model for ReynoldsAveraged Navier–Stokes Simulations of Transitional Flow", J. Fluids Eng., Volume 130, Issue 12. www.intechopen.com Transition Modelling for Turbomachinery Flows 57

24. Pacciani R., Marconcini M., Fadai-Ghotbi A., Lardeau S., Leschziner, M. A., 2009, "Calculation of High-Lift Cascades in Low Pressure Turbine Conditions Using a Three-Equation Model" ASME Turbo Expo, Orlando, FL, USA, 8-12 June, ASME paper GT2009-59557. Conf. Proc. Vol. 7: Turbomachinery, Parts A and B, pp. 433- 442. ISBN 978-0-7918-4888-3.

25. Langtry, R.B., Menter, F.R., 2009, "Correlation-Based Transition Modeling for Unstructured Parallelized Computational Fluid Dynamics Codes", AIAA J. Vol. 47, No.12,

26. Cutrone L., De Palma P., Pascazio G., and Napolitano M., 2008, "Predicting transition in two- and three-dimensional separated flows", Int. J. Heat Fluid Fl 29 504–526.

27. S. Fu and L. Wang, 2008,: "Modelling the flow transition in supersonic boundary layer with a new k–ω–γ transition/turbulence model", in: 7th

International Symposium on Engineering Turbulence Modelling and Measurements-ETMM7, Limassol, Cyprus.

28. Serdar Genec, M., Kaynak and Ü. Yapıcı, H., 2011,: "Performance of Transition Model for Predicting Low Re Aerofoil Flows without/with Single and Simultaneous Blowing and Suction" European Journal of Mechanics B/Fluids, vol 30 (2), pp. 218- 235, 2011.

29. Menter, F.R., Langtry, R.B., Likki, S.R., Suzen, Y.B., Huang, P.G., and Völker, S., 2006, "A Correlation based Transition Model using Local Variables Part 1- Model Formulation", ASME Journal of Turbomachinery, Vol. 128, Issue 3, pp. 413 – 422.

30. Langtry, R.B., Menter, F.R., Likki, S.R., Suzen, Y.B., Huang, P.G., and Völker, S., 2006, "A Correlation based Transition Model using Local Variables Part 2 - Test Cases and Industrial Applications", ASME Journal of Turbomachinery, Vol. 128, Issue 3, pp. 423 – 434.

31. Menter, F.R., Langtry, R.B. and Völker, S., 2006, "Transition Modelling for General Purpose CFD Codes", Journal of Flow, Turbulence and Combustion, Vol. 77, Numbers 1 – 4, pp. 277-303.

32. Menter, F.R., 1994, "Two-Equation eddy-viscosity turbulence models for engineering applications", AIAA Journal, Vol. 32, No. 8, pp. 1598-1605.

33. Langtry, R.B., 2006, "A Correlation-Based Transition Model using Local Variables for Unstructured Parallelized CFD codes", Doctoral Thesis, University of Stuttgart. http://elib.uni-stuttgart.de/opus/volltexte/2006/2801/

34. Fashifar, A. and Johnson, M. W., 1992, "An Improved Boundary Layer Transition Correlation", ASME Paper ASME-92-GT-245.

35. Schubauer, G.B. and Skramstad, H.K., 1948, "Laminar-boundary-layer oscillations and transition on a flat plate," NACA Rept. 909.

36. Sinclair, C. and Wells, Jr., 1967, "Effects of Freestream Turbulence on Boundary-Layer Transition," AIAA Journal, Vol. 5, No. 1, pp. 172-174.

37. Schubauer, G.B. and Klebanoff, P.S., (1955), "Contribution on the Mechanics of Boundary Layer Transition," NACA TN 3489.

38. McIlroy, H.M. and Budwig, R.S., (2005). "The Boundary Layer over Turbine Blade Models with Realistic Rough Surfaces", ASME-GT2005-68342, ASME TURBO EXPO 2005, Reno-Tahoe, Nevada, USA.

39. Zierke, W.C. and Deutsch, S., 1989, "The measurement of boundary layers on a compressor blade in cascade – Vols. 1 and 2", NASA CR 185118. www.intechopen.com 58 Low Reynolds Number Aerodynamics and Transition

40. Dorney, D.J., Lake, J.P., King, P.L. and Ashpis, D.E., 2000, "Experimental and Numerical Investigation of Losses in Low-Pressure Turbine Blade Rows," AIAA Paper AIAA- 2000-0737, Reno, NV.

41. Huang, J., Corke, T.C., Thomas, F.O., 2003, "Plasma Actuators for Separation Control of Low Pressure Turbine Blades", AIAA Paper No. AIAA-2003-1027.

42. Arts, T., Lambert de Rouvroit, M., Rutherford, A.W., 1990, "Aero-Thermal Investigation of a Highly Loaded Transonic Linear Turbine Guide Vane Cascade", von Karman Institute for Fluid Dynamics, Technical Note 174.

43. Steelant, J., and Dick, E., 2001, "Modeling of Laminar-Turbulent Transition for High Freestream Turbulence", Journal of Fluids Engineering, Vol. 123, pp. 22-30.

44. Somers, D.M., "Design and Experimental Results for the S809 Airfoil", Airfoils, Inc., State College, PA, 1989.

45. Somers, D.M., "Design and Experimental Results for the S809 Airfoil", NREL/SR-440- 6918, January 1997.

46. Langtry, R.B., Gola, J. and Menter, F.R. 2006, "Predicting 2D Airfoil and 3D Wind Turbine Rotor Performance using a Transition Model for General CFD Codes", AIAA Paper 2006-0395.

Chapter 9

A TIME EFFICIENT ADAPTIVE GRIDDING APPROACH AND IMPROVED CALIBRATIONS IN FIVE-HOLE PROBE MEASUREMENTS

Jason Town and Cengiz Camci

Turbomachinery Aero-Heat Transfer Laboratory, Department of Aerospace Engineering, The Pennsylvania State University, University Park, PA 16802, USA

ABSTRACT

Five-Hole Probes (FHP), being a dependable and accurate aerodynamic tool, are an excellent choice for measuring three-dimensional flow fields in turbomachinery. To improve spatial resolution, a subminiature FHP with a diameter of 1.68 mm is employed. High length to diameter ratio of the tubing and manual pitch and yaw calibration cause increased uncertainty. A new FHP calibrator is designed and built to reduce the uncertainty by precise, computer controlled movements and reduced calibration time. The calibrated FHP is then placed downstream of the nozzle guide vane (NGV) assembly of a low-speed, large-scale, axial flow turbine. The cold flow HP turbine stage contains 29 vanes and 36 blades. A fast and computer controllable traversing system is implemented using an adaptive grid method for the refinement of measurements in regions such as vane wake, secondary flows, and boundary layers. The current approach increases the possible number of measurement points in a two-hour period by 160%. Flow structures behind the NGV measurement plane are identified with high spatial resolution and reduced uncertainty. The automated pitch and yaw calibration and the adaptive grid approach introduced in this study are shown to be a highly effective way of measuring complex flow fields in the research turbine.

INTRODUCTION

Five-Hole Probes are used to determine the three components of the mean velocity vector, local total pressure, and local static pressure [1]. They work by selectively comparing pressure data from five ports on the probe. According to Treaster and Yocum [2], by comparing the pressure differences between these ports, flow velocity magnitude, pitch angle, yaw angle, total pressure, and static pressure can be simultaneously determined. However, this method is found to work in a range of ±30° of pitch and yaw angle. A method is suggested by Ostowari and Wentz [3] to increase the range to ±85° by using a nulling method of the probe. However, nulling is not always possible, especially in the tight clearances and rotating machinery such as the flows internal to a turbine research rig. Norwack attempts to increase the range at which a FHP may be used by developing a long, spherical probe [4]. The useable range of the probe is increased to ±65°, but the increased size makes it hard to incorporate in many cases.

Correction methods by interpolation to find the necessary coefficients have been implemented through a variety of methods. A curve fitting approach is used by Treaster and Yocum [2] and Weiz [5]. The curve fitting approach takes into account the fact that the data taken by each port is in a different location. By using an orthogonal grid, the data is curve-fitted across the measurement region and interpolated to the center port of measurement. Reichert and Wendt suggest another method of data reduction for the FHP [6]. This method replaces the pitch and yaw angles with unit vectors and develops a Taylor series based approach to find flow parameters.

Dominy and Hodson studied the effects of Reynolds number extensively. Their study showed how probe design could affect Reynolds number related errors [7]. Treaster and Yocum [2] also covered this feature and suggested that calibration should be made at the expected Reynolds number or a correction factor must be used. Methods for the detection of abnormalities in the probe are suggested by Morrison et al. [8]. These suggestions aid in the identification of probe damage and flow alignment issues.

Investigations into the effect of near-wall measurements with Five-Hole Probes were carried out by Treaster and Yocum [2] and Lee and Yoon [9]. They have concluded that measurements should be taken at least two probe diameters away from the wall. Closer distance causes blockage by the probe and acceleration of the flow, leading to greater uncertainty in the measurement. If it is necessary to operate closer than this distance, Lee and Yoon provide guidelines to make such measurements [9].

A FHP might also be used in place of laser Doppler anemometers for wake measurements, Brophy et al. [10]. The main application of the FHP would be in location where it would be difficult to use a laser, such as the rotating frame of reference in a turbine or in geometrically difficult situations to reach flow zones. Sitaram et al. have performed a detailed study on which type of probe to use within the rotating frame of a single stage compressor research rig at the Pennsylvania State University [11]. Town and Camci presented their early observations about using a subminiature Five-Hole Probe in an axial flow turbine rig in [12]. There are many recent studies on the development of FHP based aerodynamic measurement systems. Pisasale and Ahmed [13] presented a theoretical calibration approach for a FHP for highly three-dimensional flows. They also worked on a novel method for extending the calibration range of FHPs for highly three-dimensional flows [14]. Development of a functional relationship between port pressures and flow properties for the calibration and application of a multihole probe to highly three-dimensional flow was a topic of investigation in their 2004 paper [15]. Multihole probes can also be used in the determination of skin friction coefficient in turbulent flows, Lien and Ahmed [16]. The present paper presents significant improvements in FHP based aerodynamic measurements in four significant areas. The specific approach reduces the elapsed calibration time of a typical Five-Hole Probe from 3 hours down to 65 minutes for a (9×9) carpet map of pitch and yaw coefficients because of the unique properties of the new computer controlled calibration mechanism. A second major improvement is in the spatial resolution of measurements in selected high gradient areas such as the boundary layers, wakes, tip vortices, and secondary flow dominated flow zones. The third important property of the present approach is in the improved accuracy of the measurements because of an improved calibration system, the use of more accurate positioning of the probe, the use of highly improved present day transducers, and a careful selection of tubing. Finally, the current approach reduces turbine facility run-time significantly. The new system increases the number of data points that can be collected in a two-hour period from 366 points to 868 points, an increase of 160%.

SYMBOLS

AFTRF: Axial Flow Turbine Research Facility:

$C_{P,pitch}$ Pitch coefficient

$$\frac{(P_5 - P_4)}{(P_1 - \overline{P})} \tag{1}$$

$C_{P,static}$: Static pressure coefficient

$$\frac{\left(\overline{P} - P_{\text{static}}\right)}{\left(P_1 - \overline{P}\right)} \tag{2}$$

$C_{P,\text{total}}$: Total pressure coefficient

$$\frac{\left(P_1 - P_{\text{total}}\right)}{\left(P_1 - \overline{P}\right)} \tag{3}$$

$C_{P,\text{yaw}}$: Yaw coefficient DAQ

$$\frac{\left(P_2 - P_3\right)}{\left(P_1 - \overline{P}\right)} \tag{4}$$

DAQ: Data acquisition device

$\overline{P}$: Average value of outside pressure points (Pa)

$$\frac{\left(P_2 + P_3 + P_4 + P_5\right)}{(4)} \tag{5}$$

P_1: Pressure point 1, Figure 1
P_2: Pressure point 2, Figure 1
P_3: Pressure point 3, Figure 1
P_4: Pressure point 4, Figure 1
P_5: Pressure point 5, Figure 1
P_S: Static pressure at defined location (Pa)
P_T: Total pressure at defined location (Pa)
P_{static}: Local FHP measured static pressure (Pa)
P_{total}: Local FHP measured total pressure (Pa)
V_1: Transducer voltage measured at point 1, Figure 1
V_2: Transducer voltage measured at point 2, Figure 1
V_3: Transducer voltage measured at point 3, Figure 1
V_4: Transducer voltage measured at point 4, Figure 1
V_5: Transducer voltage measured at point 5, Figure 1
V: Absolute velocity magnitude (m/s)
c: Calibration constant (Pa/V)
exit: Measurement location downstream of rotor
inlet: Measurement location upstream of NGV
local: Measurement location in intraspace at specific point
r: Radius to probe traverser track
u: Velocity component in the probe relative x-direction (m/s)
v: Velocity component in the probe relative y-direction (m/s)
w: Velocity component in the probe relative z-direction (m/s)
x: Distance of traverser from probe parallel to ground (zero) location
z: Zero voltage measurement (V)
α: Probe relative pitch angle (°)
β: Probe relative yaw angle (°)
θ: Desired measurement azimuthal angle
ρ: Density (kg/m³).

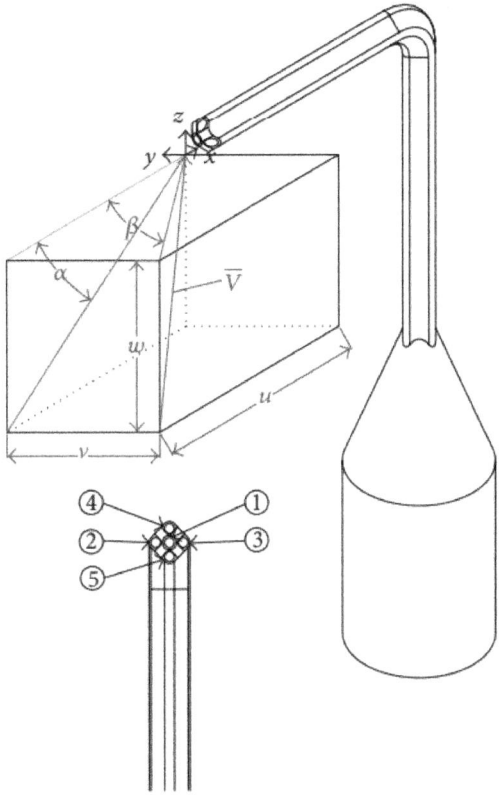

Figure 1: Isometric view of a subminiature Five-Hole Probe with velocity vector, positive angles, and positive velocity components and probe port number assignments.

MATERIAL AND METHODS

Objectives

First of the specific objectives is the reduction of elapsed calibration time of a typical FHP for use in the AFTRF from three hours down to about an hour for a (9×9) carpet map of pitch and yaw coefficient. A second major objective is in the improvements in the spatial resolution of measurements in selected high gradient areas such as boundary layers, wakes, tip vortices, and secondary flow dominated zones. The third important objective of the present approach is in the improved accuracy of the measurements. The final objective is about significantly reducing the AFTRF turbine facility run-time for a selected FHP measurement effort.

Calibration Hardware

A flowchart for the acquisition of calibration data is shown in Figure 2. The commands, interface, and data logging at the computer level are written in LabVIEW. The 16-bit A/D converter system (DAQ) is made by Measurement Computing Corporation (MCC model USB-1608FS). It can obtain 200 k samples per second, interfaces through USB, and has an accuracy of ±0.68 mV at an input range of ±1 V. The 16-bit capabilities are essential to help reduce measurement uncertainty during the calibration and measurements.

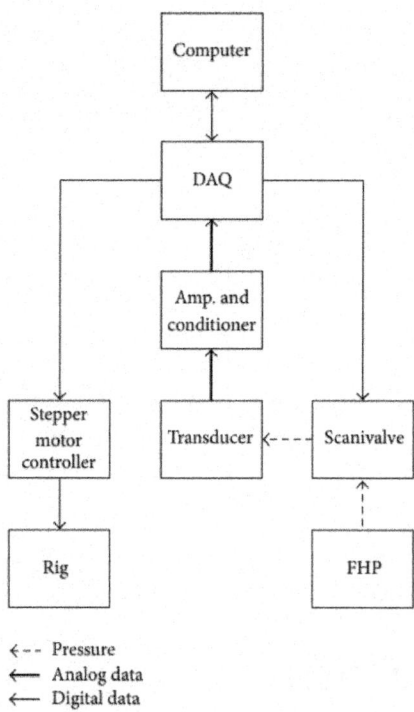

Figure 2: Calibration block diagram for Five-Hole Probe measurements.

The pressure is measured by a Validyne DP15 low pressure transducer with a 3500 Pa diaphragm. The transducer's accuracy is rated at 0.25%, (±3.5 Pascal) of the full-scale measurement. Only one transducer is used to measure all five ports; its reference port is left open to atmosphere, and efforts are made to isolate the transducer thermally, electronically, and mechanically. Using one transducer for all five ports eliminates the bias errors coming from individual transducer zero values.

The transducer is connected to a Scanivalve Corp. 48-channel mechanical pressure selector. The specific electrical commands to step and reset the scanner are provided by the digital output D/A of the DAQ. The mechanical scanning approach with one transducer reduces measurement uncertainty by canceling out any thermal shift and calibration error the transducer might measure. It also reduces the total cost of the system but comes up with increased measurement time for the calibrations and measurements.

Previously employed methods of pitch and yaw calibration were applied by hand. To increase the accuracy of movements and decrease the total time taken for a complete calibration map, computer actuated rotary tables for pitch and yaw movements are used. Two different rotary tables are used, both provided by Velmex Inc. The larger of the two, model B4800TS, is used to change the pitch angle. For each half step it moves 0.0125°. A second table, model B5990TS, is used to for yaw adjustments. For each half step it moves 0.010°.

The calibration rig shown in Figure 3 is designed to minimize movement of the probe tip in the direction normal to the outlet of the wind tunnel. The design intent is to keep the probe tip as near the intersection of the pitch and yaw axes as possible. This reduces the arm length of the probe and its displacement within the calibration jet.

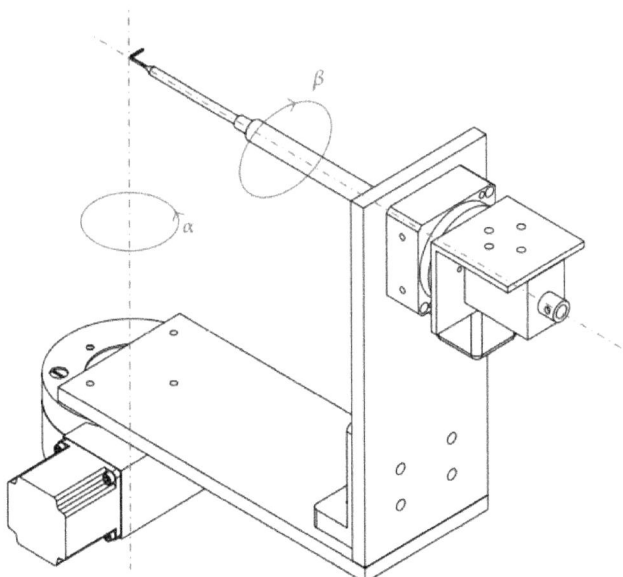

Figure 3: Calibrator design with pitch (α) and yaw (β) movements.

A subminiature FHP shown in Figure 1 is designed and created on site using five hypodermic needles. It has a square cross-section with one hole in the center surrounded by four others located above, below, left, and right of the center. All sides of the probe have been beveled to a 45° angle, leaving the center port in a normal plane to the flow. The specific hypodermic needles reduce the size of the probe to a maximum diameter of 1.68 mm at the tip. The inherently small diameter of hypodermic tubing and the long length of tubing lead to a very large length to diameter ratio. Subminiature FHPs usually require relatively long settling and data acquisition times. However, the small tip size increases the spatial resolution of the measurements and allows the probe to be inserted into the complex internal flow areas of most turbomachinery passages. An additional Pitot probe is also used to measure the calibration tunnel's free jet axial velocity, total pressure, and static pressure.

The calibration facility, shown in Figure 4, consists of an open loop wind tunnel with an axial blower, a diffused housing with multiple screens, a plenum chamber, a high area ratio circular nozzle, a circular to square transition nozzle, and a section of constant cross-sectional duct. The compressor is 45.7 cm in diameter and is driven by a variable speed motor rated up to 7.5 kW. The tests are performed in the free jet just outside the constant cross-sectional duct. Free jet velocities are continuously adjustable via an AC inverter up to 28 m/s. Turbulent flow characteristics in the test section can be adjusted to turbulence intensity values between 0.5% and 1.2% by the use of calibrated screens and biplane turbulence promoters. Details of the test section flow quality can be found in Wiedner [1], Kuisoon et al. [17, 18], and Camci and Rizzo [19].

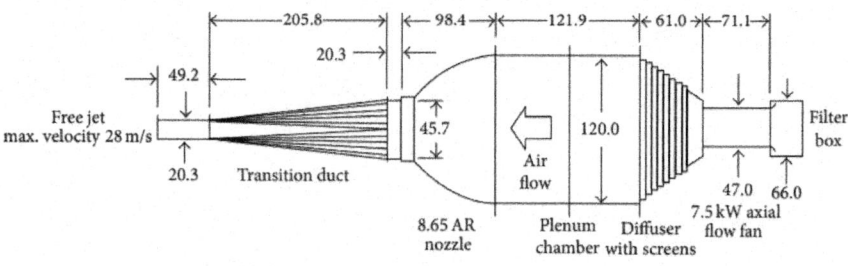

Figure 4: Calibration wind tunnel, dimensions in centimeters (not to scale).

Calibration Technique

A modified version of the calibration/reduction technique of a nonnulling FHP used by Treaster and Yocum [2] is used in this paper. The main difference between the two methods is in the way pitch and yaw angles are defined.

Figure 1 defines the positions of the holes, the coordinate system for the probe, the positive angles of the probe, and the positive velocity components. In this configuration, being positioned on the probe facing the incoming flow, a positive pitch value would occur when the flow was coming from below the probe (nose up). Positive yaw value would occur when the flow is coming from the left (nose right). The only manual input into the calibration sequence is the initial hand positioning of the probe at zero pitch, zero yaw angles. The calibration grid uses an improved 81-point (9 × 9) configuration with more data points in the nonlinear region near the maximum acceptance range of the probe (±30°). Calibration maps for coefficients of yaw, pitch, total pressure, and static pressure are calculated directly from the measurements of the FHP and Pitot probe as shown in Figures 8, 9, and 10. In most earlier studies, the carpet maps were limited to a 49-point (7 × 7) configuration or less. One of the main concerns with calibrating a FHP is producing a high quality map with reduced absolute error. One concern is about instant changes in calibration flow quality that can be caused by laboratory disturbances or unwanted air currents. Another source of error can occur when the probe is first aligned with the flow. The probe is initially aligned by hand and is prone to human error. Only one transducer is used to sample pressure from all pressure measurement points. Five input channels to the rotary pressure scanner are required for the FHP ports and two input channels are needed for the Pitot probe documenting the total pressure and static pressure in the test section.

Using a single transducer for all seven pressure measurements during calibration measurably increased the elapsed time for a calibration. However, this approach significantly reduces calibration error. Equation (6) states that pressure is a function of the measured voltage V_n, the zero z, and a calibration factor c. Since the zero and the calibration factor are for only one transducer, they can be considered constant for all pressure measurements. The analysis will only take $C_{P,\text{total}}$ from (3) into consideration, though it can be done with any of the other pressure coefficient equations. Substitution of (6) into (3) results in (7). Since c is in every term, canceling it out leads to (8). The result of canceling out the z term in the numerator and simplifying the z term in the denominator is shown in (9). Finally, the z term in the denominator is cancelled out and the result is shown in (10). The result shows that when calculating $C_{P,\text{total},}$ or for that matter any of the C_p values, the zero and calibration factors cancel out. Thus, using one transducer eliminates the source of error that could be caused by a typical calibration of a transducer and incorrect zeroing procedure. Consider

$$P_n = \left(V_n - z\right)c,$$

$$(6)$$

$$C_{P,\text{total}}$$
$$= ((V_1 - z)c - (V_T - z)c)$$
$$\times \Bigg((V_1 - z)c$$
$$- \frac{(V_2 - z)c + (V_3 - z)c + (V_4 - z)c + (V_5 - z)c}{4} \Bigg)^{-1},$$

$$(7)$$

$$C_{P,\text{total}}$$
$$= ((V_1 - z) - (V_T - z))$$
$$\times \Bigg((V_1 - z)$$
$$- \frac{(V_2 - z) + (V_3 - z) + (V_4 - z) + (V_5 - z)}{4} \Bigg)^{-1},$$

$$(8)$$

$$C_{P,\text{total}} = \frac{V_1 - V_T}{(V_1 - z) - (V_2 + V_3 + V_4 + V_5 - 4z)/4},$$

$$(9)$$

$$C_{P,\text{total}} = \frac{V_1 - V_T}{V_1 - (V_2 + V_3 + V_4 + V_5)/4}.$$

$$(10)$$

Calculating Unknown Flow Variables

A FORTRAN code developed in-house is the traditional method of data reduction in the measurements of a FHP. The recent fully automated LabVIEW implementation of the same analytical calibration/reduction procedure is shown to produce identical results or better when compared to our past manual calibration/reduction system. Input to the code requires the current ambient temperature (T), absolute static pressure (P_s), and the five pressures as measurements of the FHP. The program then determines the pitch angle α through a series of linear interpolations. The interpolation scheme calls $C_{P,\text{pitch}}$ to be calculated for each possible pitch angle in the 81- point carpet map where there are nine possible pitch angles at a constant $C_{P,\text{yaw}}$. A second interpolation calculates the value of pitch angle by using the measured value of $C_{P,\text{pitch}}$. Yaw angle β is calculated in a similar interpolation scheme. This interpolation scheme calls $C_{P,\text{yaw}}$ to be calculated for each possible yaw angle at a constant $C_{P,\text{pitch}}$. The next interpolation calculates the exact value of yaw angle by using the measured value of $C_{P,\text{yaw}}$. A calibration chart used to visualize the data of the interpolations of pitch and yaw angle can be found in Figure 8.

Pitch and yaw angles are used for the interpolation of $C_{P,\text{total}}$ and $C_{P,\text{static}}$. The interpolation works similarly to pitch and yaw interpolation. First, α is held constant and an array of $C_{P,\text{total}}$ or $C_{P,\text{static}}$ is found with an array of β. The previously found value of β is then used to calculate $C_{P,\text{total}}$ and $C_{P,\text{static}}$, respectively. The charts used for the calculation of $C_{P,\text{total}}$ and $C_{P,\text{static}}$ can be found in Figures 9 and 10. Hence,

$$P_{\text{total}} = P_1 - C_{P,\text{total}} \left(P_1 - \overline{P} \right), \tag{11}$$

$$P_{\text{static}} = \overline{P} - C_{P,\text{static}} \left(P_1 - \overline{P} \right), \tag{12}$$

$$V = \left[\frac{2 \left(P_T - P_S \right)}{\rho} \right]^{1/2}, \tag{13}$$

$$u = V \cdot \cos \alpha \cdot \cos \beta, \tag{14}$$

$$v = V \cdot \sin \beta, \tag{15}$$

$$w = V \cdot \sin \alpha \cdot \cos \beta. \tag{16}$$

Axial Flow Turbine Research Facility

The Axial Flow Turbine Research Facility at Pennsylvania State University currently consists of a single stage state-of-the-art HP turbine with 29 nozzle guide vanes (NGV) and 36 rotor blades. Figure 5 shows a cross-sectional view of the AFTRF. Flow enters through the inlet bellmouth and accelerates through the NGV. A detailed explanation of the design and characteristics of the AFTRF can be found in Lakshminarayana et al. [20].

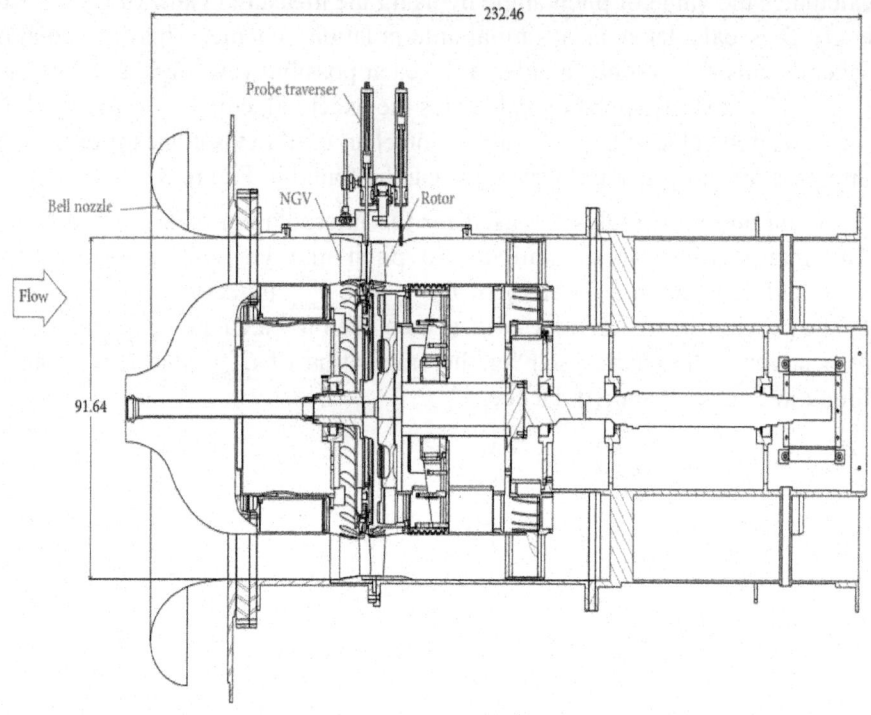

Figure 5: Axial Flow Turbine Research Facility (AFTRF) cutaway (dimensions in cm).

The recent probe traverser as shown in Figures 5 and 6 is a modification of a previous design used in the AFTRF. In the previous design, circumferential movements were achieved by a belt system. The flexible belt system had noticeable play in it, and the new system replaces the belts with a precision built linear traverser shown in red. Replacing the belt system with a traverser greatly increases the accuracy and reduces the play of the system. There are two radial traversers that are connected to the trolley (light blue), and the probe holders are shown in yellow. It is designed with two radial traversers so that the system could take measurements in locations in between the NGV and rotor (intraspace), or downstream of the rotor. The new traversing system also includes improvements to the stepper motor drivers, which helps the traverser to move much faster than previously.

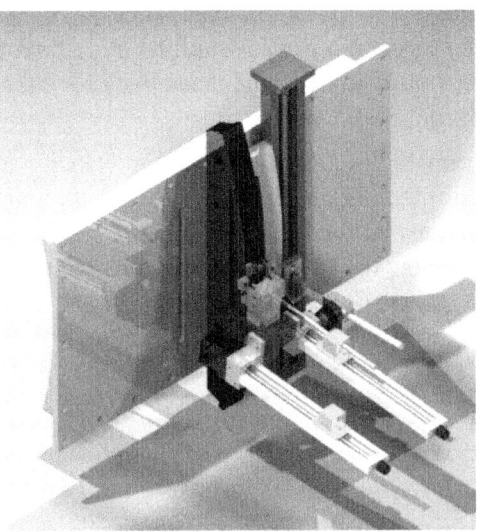

Figure 6: Probe traverser attached to the AFTRF instrumentation window, isometric view.

A cylindrical coordinate system is used within the AFTRF. Movements of the trolley could be translated to azimuthal position by calculation of arc length. Equation (17) is the result of the calculation to find the distance the traverser must move to change to a particular azimuthal angle θ. All distances are derived from a zero angle; the probe is parallel to the normal of the traverser. The radius r is the distance to the surface of the track that the trolley rides upon. The final variable, x, is the distance that the traverser must move from the zero position. To move from 1° to 2°, x is calculated for both cases and the values are subtracted to find the correct distance:

$$x = \frac{r \cdot \tan \theta}{\sqrt{\tan^2\theta + 1}}. \tag{17}$$

Although a conventional Scanivalve mechanical pressure scanner was used in the calibration of the FHP measurements for the improved calibration accuracy, our current AFTRF FHP measurements are performed using state-of-the-art electronic pressure scanners. Thirty-two-channel ZOC22b units from Scanivalve Corporation were employed in most of the turbine runs for faster operation and better thermal stability of the measurements.

Calculation of density in our FHP based calibrations and actual measurements require a measurement of local temperature. A K type thermocouple based probe is employed in our temperature measurements throughout the AFTRF. The thermocouple signal is referenced/amplified and converted into engineering units in a custom NI-9213 thermocouple processing unit. Our final temperature measurement accuracy for calibrations and AFTRF is estimated to be around ±0.15°C.

Fluid flow through the AFTRF varies in velocity to the extent that a probe calibrated for the inlet could not be used in the intraspace or rotor exit relative frame locations but could be used in the rotor exit absolute frame of reference. Table 1 covers the values of midspan velocity in meters per second. Present measurements are performed in the intraspace location, so the probe is calibrated at about 66 m/s.

Table 1: Predicted midspan flow speeds and locations

Location	Inlet	Intraspace	Absolute exit	Relative exit
Midspan velocity (m/s)	16.15	66.30	20.05	66.88

RESULTS

Five-Hole Probe Calibration

The carpet map as shown in Figure 7 was obtained four times in subsequent runs in an effort to establish the repeatability of the calibration process. The center of each cross represents the average value, while the four points surrounding each cross represent the data collected from each run. Nearly all points within the ±20° range have good grouping and are close to the average value. In the outlying regions of the calibration, those greater than ±20°, the grouping is not as tight and initial alignment errors are exacerbated. The star shaped carpet map shown in Figure 7 is not perfectly symmetrical because a dimensionally perfect and symmetrical FHP is very difficult to manufacture because of the probe's small size and the inherent machining imperfections.

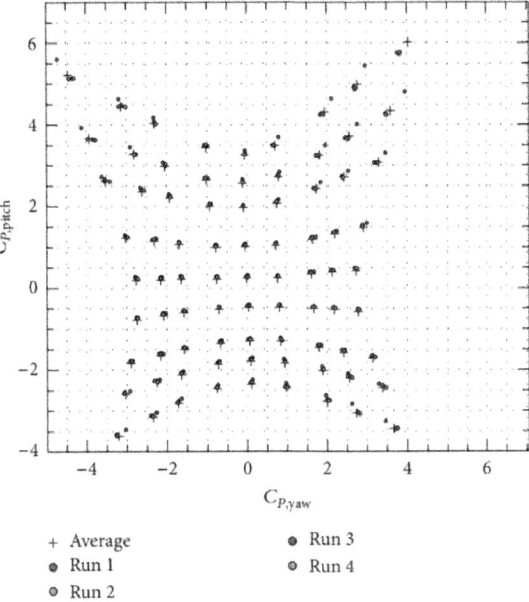

Figure 7: Average values (crosses) with data spread of four runs (points).

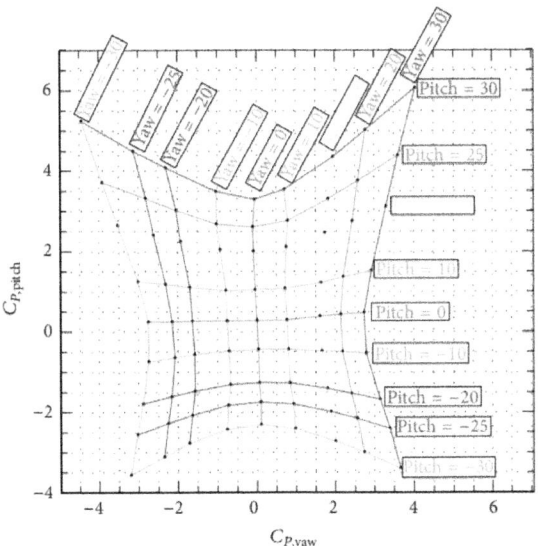

Figure 8: Coefficient of yaw and pitch with lines of constant pitch or yaw angles, 81 (9 × 9) points.

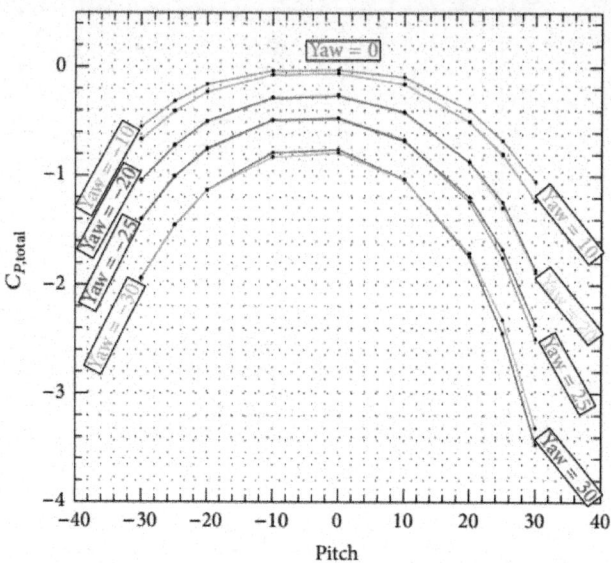

Figure 9: Pitch and coefficient of total pressure with lines of constant yaw, 81 (9 × 9) points.

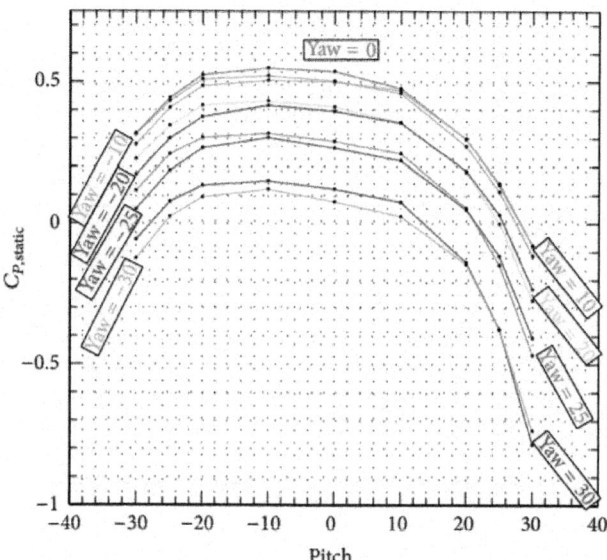

Figure 10: Pitch and coefficient of static pressure with lines of constant yaw, 81 (9 × 9) points.

Calculations of $C_{P,\text{pitch}}$ and $C_{P,\text{yaw}}$ are found directly through (4) and (1). The results are used to interpolate pitch and yaw angle values with the help of the data presented in Figure 8 that is a typical averaged carpet map produced by the current automated calibration approach. Figure 9 presents the variation in $C_{P,\text{total}}$ with respect to pitch and yaw angle. Interpolation is done to find $C_{P,\text{total}}$ using the pitch and yaw found in the previous section. Similarly, Figure 10 provides $C_{P,\text{static}}$ as a function of pitch and yaw. $C_{P,\text{static}}$ can be recovered from the FHP measurements using the previously found pitch and yaw angles.

Previous design of calibration called for the probe to be moved by hand. One of the major advantages of the current computer automated system is stepper motor driven movements. No longer relying on human movement allows for a much greater degree of accuracy. The stepper motor controller can move the turntables in steps as fine as 0.0125°. Once the initial zero pitch and zero yaw position are defined at the start of each run, the computer driven mechanism can move to a new position with excellent spatial resolution and accuracy.

Calibrations of a FHP by manual pitch and yaw angle adjustments are long and arduous tasks. Previously, a 49-point map in a manual calibration effort took at least three hours to complete since a high quality adjustment of each pitch and yaw angle required great care. When the design was changed to the current automated system, a 49-point map took 65 minutes, and an 81-point map took 100 minutes. Further reduction in elapsed time is seen if multiple transducers are used, at the cost of previously mentioned accuracy improvements with the single transducer approach. An 81-point grid with multiple transducers takes approximately 22 minutes to complete. Ensemble averaging from four individually obtained carpet maps can also be very easily obtained in the current computer driven system in a time efficient manner. Ensemble averaging is an excellent way of removing some of the error originating from initial alignment of the probe. Paying attention to properly recording the transducer zero voltages just before a turbine run starts is an effective way of improving accuracy in the multiple transducer approach.

The method of attachment of a FHP to the pitch and yaw calibrator is extremely critical. The style and quality of the mechanical attachment influence the movement of the probe and the value of the calibration. Large swings and displacements can produce large errors in the calibration from spatial nonuniformity, turbulence decay, or shear layer mixing. The calibrator in Figure 3 shows a design where the probe's tip is located near the intersection of the pitch and yaw rotational axes. The improved design makes the calibration more accurate and reduces the uncertainties.

Due to the nature of the design of the probe, the initial alignment must be performed by hand. This is because of the small size and unforeseen defects that make nulling the probe unfeasible. However, a few techniques are developed in order to increase the initial alignment accuracy. First, a plumb bob is used to align the calibrator base so that it is parallel to the exit of the wind tunnel. This helps to insure that the probe will be held normal to the exit of the wind tunnel and parallel to the streamlines in the pitch direction. Yaw angle alignment is done with the help of a visible, horizontal laser beam. Placing a piece of paper over the exit of the tunnel allows for comparison of the shape of the shadow of the probe. The shadow is then brought to its minimum size by making small adjustments with the yaw stepper motor. These two alignments are done at the beginning of every run.

An uncertainty analysis is prepared for the total pressure, the static pressure, and velocity as defined by (11), (12), and (13). An adaptation of a method set forth by Taylor [21] is used. Equation (18) is an example of the total pressure uncertainty estimates. The other variables follow the same processes but are not presented for brevity:

$$
\begin{aligned}
\delta P_T &= \left[\left(\delta P_1 \frac{\delta P_T}{\delta P_1}\right)^2 + \left(\delta P_2 \frac{\delta P_T}{\delta P_2}\right)^2 + \left(\delta P_3 \frac{\delta P_T}{\delta P_3}\right)^2 \right. \\
&\left. + \left(\delta P_4 \frac{\delta P_T}{\delta P_4}\right)^2 + \left(\delta P_5 \frac{\delta P_T}{\delta P_5}\right)^2 \right. \\
&\left. + \left(\delta C_{P,total} \frac{\delta P_T}{\delta C_{P,total}}\right)^2\right]^{1/2},
\end{aligned}
$$

$$\delta P_1 = \delta P_2 = \delta P_3 = \delta P_4 = \delta P_5 = \pm 5 \text{ Pa},$$

$$\delta C_{P,total} = \pm \frac{1}{2}\left(C_{P,total\ max} - C_{P,total\ min}\right).$$

$$(18)$$

$$(19)$$

$$(20)$$

Accuracy of the transducer is approximately ± 3.5 Pa. Due to additional equipment and the high length to diameter ratio of connecting tubes, the accuracy decreased to ± 5 Pa. This is expressed in (19). The uncertainty of the total pressure coefficient needs to be estimated using (20).

Total pressure uncertainty is shown in Figure 11 where the most accurate region of the measurement is located slightly to the left of the zero pitch, zero yaw location. This is due to small defects that are asymmetries in the probe tip shape. The $30°$ pitch, $-30°$ yaw location sees a much larger variation in total pressure. The dashed box represents a $\pm 20°$ region that is to help identify the range for which the probe is most accurate. When taking measurements, the incidence angle of the flow to the probe is kept within this region.

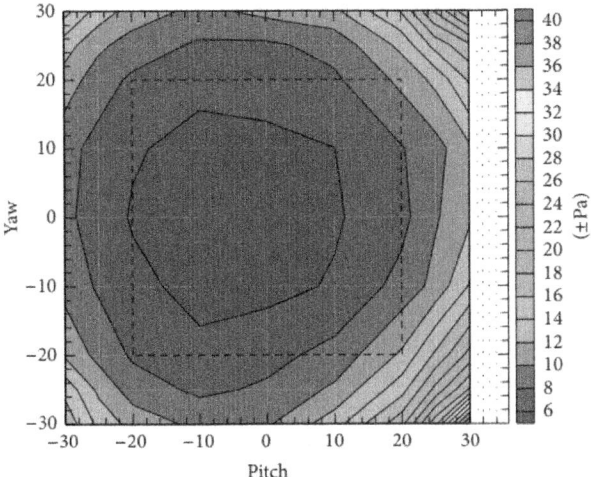

Figure 11: Total pressure uncertainty at 20 m/s, 81 (9 × 9) points (in terms of ±Pa).

Static pressure uncertainty at 20 m/s is shown in Figure 12. The uncertainty results are not symmetric due to imperfections in the probe. The uncertainty near the center is larger. This is most likely because of the way the static pressure is calculated. The probe measures greater pressure when it is aligned to flow leading to an increase in uncertainty through greater measurement variability. When the probe is taking measurements near the maximum range, some of the pressure measurements decrease and the overall uncertainty is reduced.

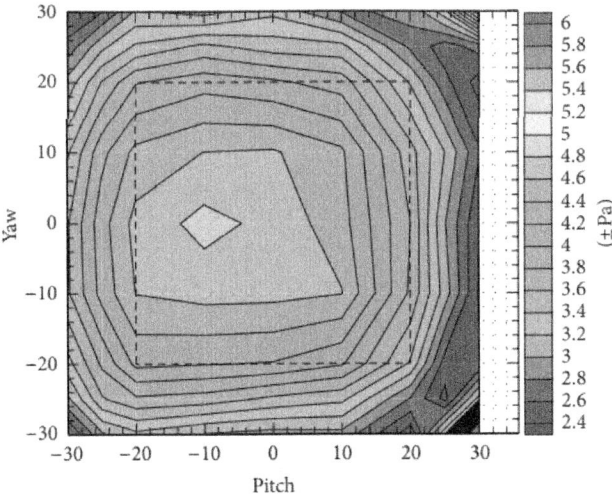

Figure 12: Static pressure uncertainty at 20 m/s, 81 (9 × 9) points (in terms of ±Pa).

Velocity uncertainty shown in Figure 13 has a minimum to the left of center. It reaches a maximum uncertainty greater than ±2.5 m/s at pitch 30° and yaw −30°. Uncertainty in the recommended ±20° range is below ±0.8 m/s.

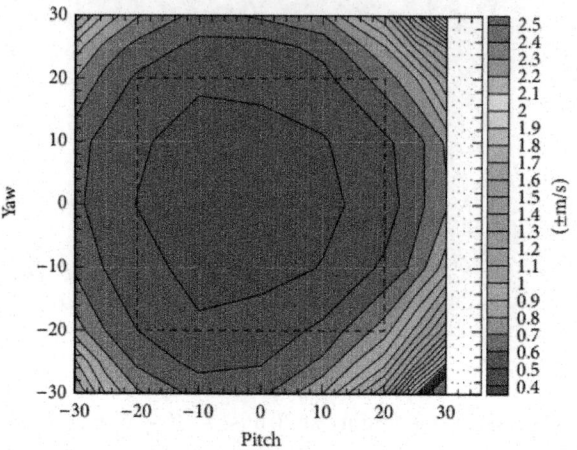

Figure 13: Velocity uncertainty at 20 m/s, 81 (9 × 9) points (in terms of ±m/s).

Figure 14 represents the relative velocity uncertainty. The smallest error is located to the left of center in the map, while the largest error is located near the edges. At pitch 30°, yaw −30°, the error is greater than 10%. Within the dashed box range of ±20° no error is found to be greater than 4%. The estimated error in the range of ±10° is about 2% of the calibration tunnel velocity.

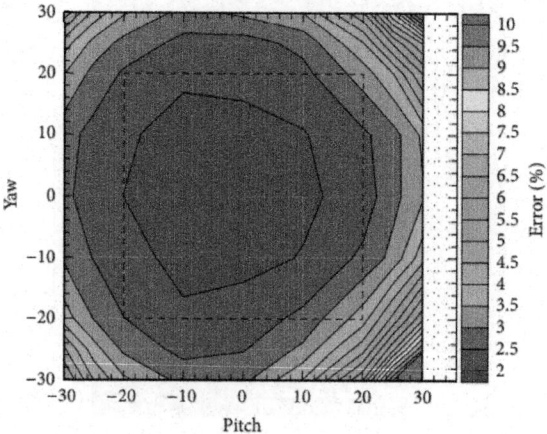

Figure 14: Velocity percent error at 20 m/s, 81 (9 × 9) points (in terms of percentage from baseline).

The preceding figures determine that the most ideal operation of this probe would be within the ±20° range that is defined by the dashed box. A pitch and yaw range of ±30° is also possible. However, the elevated measurement uncertainty should be carefully evaluated in this range.

AFTRF Intraspace Five-Hole Probe Measurement Results

The new traverser greatly reduced the amount of time it takes to move from one point to another. The maximum allowable testing period for the AFTRF was about two hours. In older previous experiments performed in the AFTRF, a 336-point mesh was used with the same sampling time for each measurement point. This mesh would take approximately two hours to complete and could not handle adaptive gridding. Improvements in the traversing system increased the number of points that are measured in a two-hour period to 868. The adaptive mesh and traverser improvements increased the number of points taken by nearly 160% over the previously used meshing techniques for the same two-hour test.

The addition of the adaptive grid approach into the mesh generation allows fine measurements in locations of greater interest and large gradients. The iterative process that is done for the adaptive gridding is currently manual. The initial grid of 165 points is shown in Figure 15. This grid is refined in areas of wake, boundary layer, and endwall vortices until the grid becomes what is shown in Figure 16. The refined grid of Figure 16allows for less time to be spent in regions of less variation and more time spent in large gradient areas. Consider

$$C_{P,\text{total}} = \frac{P_{T,\text{local}} - P_{T,\text{inlet}}}{(1/2)\,\rho V_{\text{inlet}}^2},$$
(21)

$$C_{P,\text{static}} = \frac{P_{S,\text{local}} - P_{T,\text{inlet}}}{(1/2)\,\rho V_{\text{inlet}}^2}.$$
(22)

Coefficient of total pressure as defined by (21) for one vane pitch (12.41°) is shown in Figure 17. In this figure, the more negative values "going toward blue" indicate greater total pressure loss. The boundary layer near the hub is very small. The probe at the zero span location is currently behind a backward facing step which reduces size of the boundary layer. The passage vortex of the hub centered in the blue region is in the range of, θ, from −1° to 1°, and span 0.05 to 0.10. The vane wake is the yellow region of the total pressure loss that curves through the measurement plane. Above 90% span, the uncertainty is increased as the probe is nearing the slot. This slot is used to access the intraspace measurement plane and the boundary layer can be seen along with the casing passage vortex.

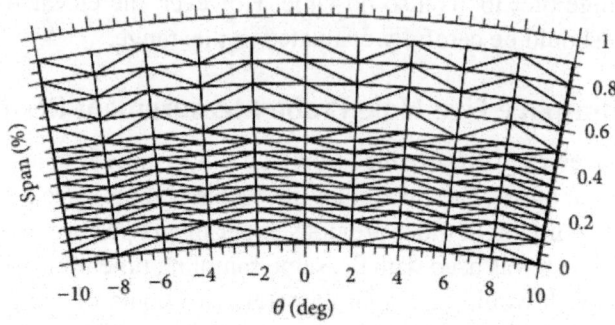

Figure 15: Initial measurement grid, 165 points.

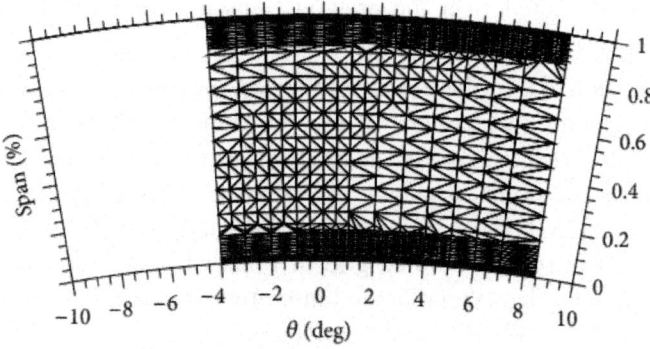

Figure 16: Refined measurement grid, 868 points.

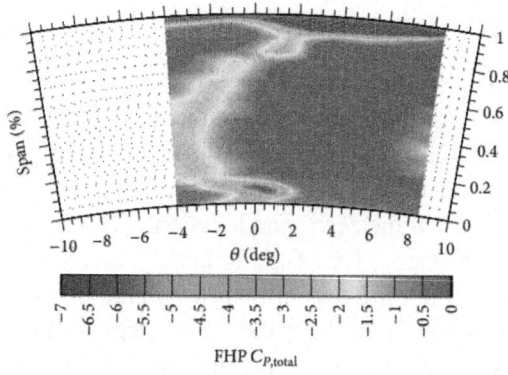

Figure 17: FHP coefficient of total pressure, intraspace.

Coefficient of static pressure measured by the FHP in the intraspace of the AFTRF as defined by (22) is shown in Figure 18. One can see the effects the secondary flows in Figures 17 and 18. The highest static pressure can be found near the hub; the lowest static pressure is found near the casing. The casing has greater uncertainty because the measurements are taken near the instrumentation slot. Static pressure over the midspan varies from being values of −23 to values of −28.

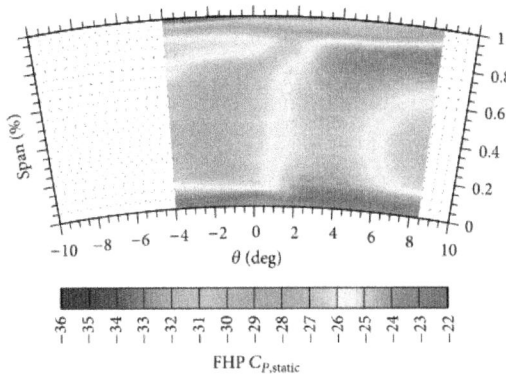

Figure 18: FHP coefficient of static pressure, intraspace.

Velocity magnitude measured by the FHP is shown in Figure 19. The highest velocity is found near the hub and velocity is reduced near the casing. The blue region in Figure 17 identified as the hub passage vortex appears here as a yellow deficit zone in velocity in the same location. Inspection can make out the outline of the vane wake, but it is not as clear as the hub passage vortex. Velocities are the lowest near the casing. This is an area of increased uncertainty due to its proximity to the instrumentation slot. The casing passage vortex and boundary layer can be identified.

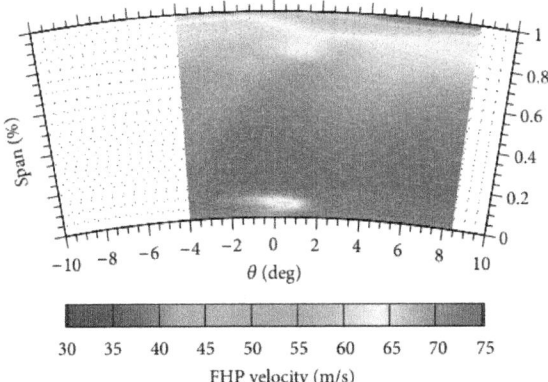

Figure 19: FHP velocity measurement, intraspace.

Since local velocity is known at each measurement location, mass averaging can be performed with the FHP results. This is achieved by taking the entire measurement area, partitioning it into slices of constant radius, and then finding the mass average of each slice of constant radius. Mass averaged total and static pressure are reported in Figure 20. The effect of the hub passage vortex is seen in the five to ten percent span region causing a higher static pressure and a lower total pressure. For much of the span the total pressure remains near zero, while static pressure is becoming greater as it moves toward the casing. Near the casing total and static pressure measurements converge due to loss from the casing boundary layer, passage vortex, and the increased uncertainty of the instrumentation slot.

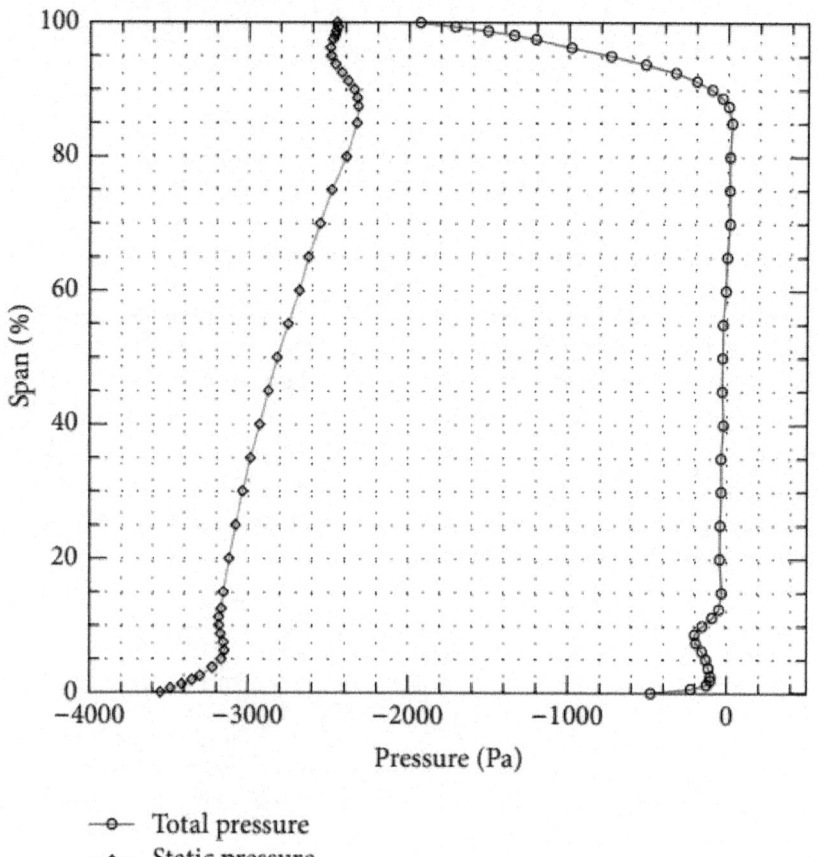

-o- Total pressure
-o- Static pressure

Figure 20: Mass averaged FHP measured total and static pressure, intraspace.

Figure 21 shows the pitch angle measured by the FHP. In this case, a positive pitch angle means the flow is moving radially toward the center of the test rig, or from the casing to the hub. Here, the effects of the backward facing step can be seen as an increased alpha near the hub. The entire passage shows the flow is moving from the casing toward the hub as shown in Figure 20. Interference from the instrumentation slot results in flow moving radially toward the hub near 100% span.

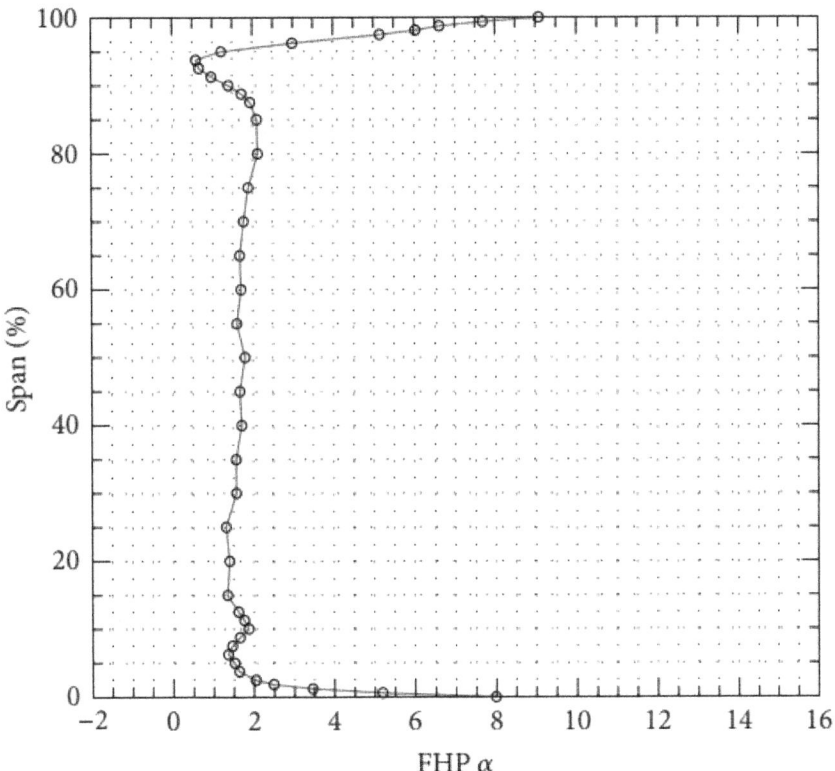

Figure 21: Mass averaged FHP measured pitch (α) angle, intraspace.

Mass averaged yaw angle is shown in Figure 22. The effect of the hub passage ortex is seen near 5% span as a quick increase and decrease in yaw angle. Near 100% span, a yaw angle decrease is seen from the casing boundary layer and casing passage vortex. Increased uncertainty in this region is due to the instrumentation slot.

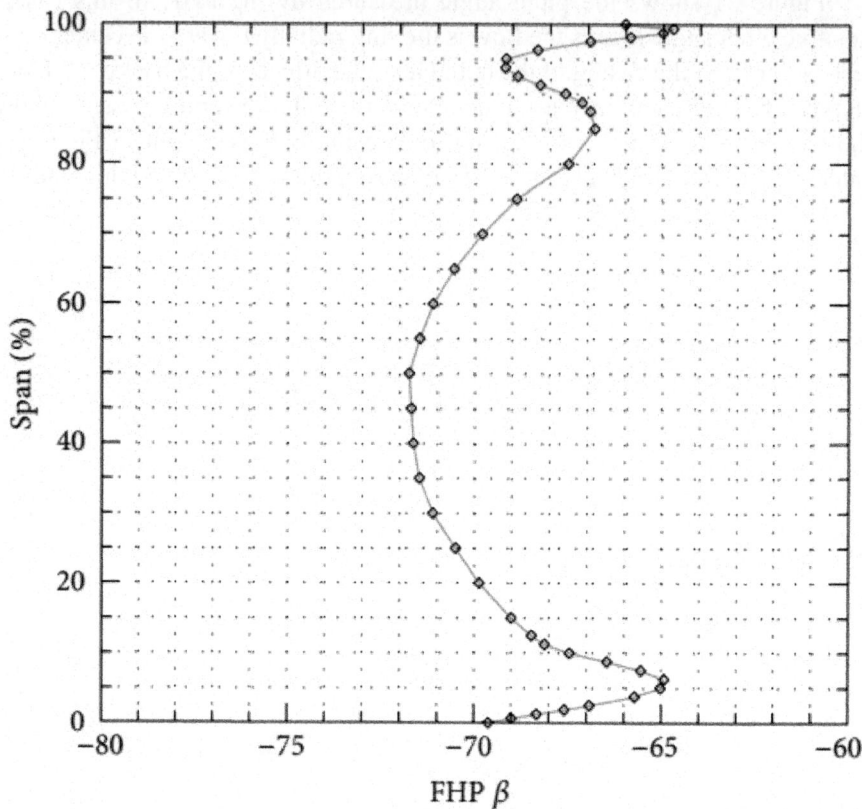

Figure 22: Mass averaged FHP measured yaw (β) angle, intraspace.

The mass averaged components of velocity are shown in Figure 23. Azimuthal velocity β is the largest component of velocity. The fluid out of the NGV is highly swirled. A reduction in this velocity from the hub passage vortex can be seen from five to ten percent. Velocity near the casing boundary layer, casing passage vortex, and instrumentation slot locally reduces the azimuthal component of velocity. A nearly constant and very small magnitude radial velocity is shown over the entire span, signifying the flow is moving toward the hub from the casing. Axial velocity is the highest near the hub passage vortex, slowly changes over the whole span, and then quickly drops in the presence of the casing boundary layer and instrumentation slot. The velocity magnitude closely follows the azimuthal component. The effects of the hub passage vortex can be found from five to ten percent, and the casing effects are also shown as a reduction in velocity.

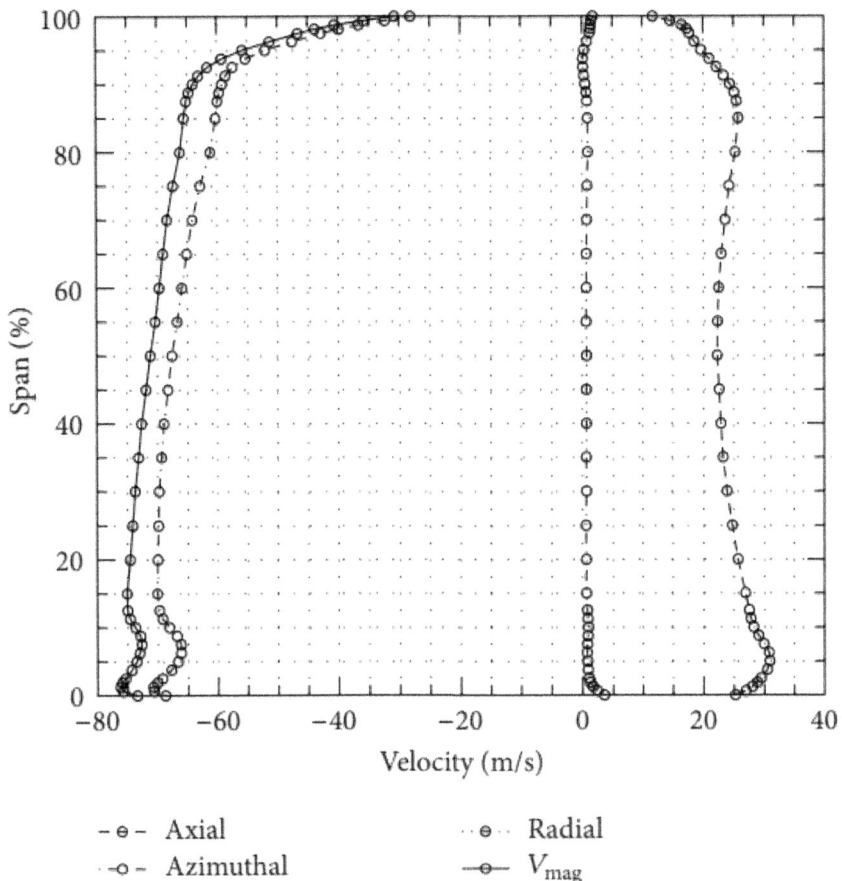

Figure 23: Mass averaged FHP measured velocity components, intraspace.

CONCLUSIONS

A subminiature Five-Hole Probe is calibrated using a newly designed automated pitch and yaw calibration system and an overview of various recent FHP implementations is given. The automated system reduces the amount of time for an 81-point (9 × 9) carpet map calibration from three hours to 65 minutes.

Ensemble averaging of multiple runs measurably reduces the uncertainty of FHP calibrations in carpet maps. Uncertainty is further reduced by realigning the probe at a zero pitch, zero yaw position with the help of a laser level and plumb bob for each calibration run.

Using only one transducer during calibration reduces calibration uncertainty by eliminating the zero and calibration terms.

The fully automated pitch and yaw calibrator is designed to reduce uncertainty caused by streamwise displacement of the probe by keeping the tip of the probe close to the intersection of the pitch and yaw rotational axis. The pitch and yaw calibrator uses state-of-the-art mechanical components, stepper motors, indexers, and computer controls to provide incremental angular changes in reduced time duration when compared to the conventional manual calibrator.

An uncertainty analysis is completed on the FHP calibration data. Within a ±20° range total pressure uncertainty is found to be less than ±12 Pa, static pressure uncertainty is less than ±5 Pa, and velocity uncertainty is less than ±1 m/s (4%).

A calibrated probe is placed into a large-scale axial turbine research rig. The research rig AFTRF has a recently modified probe traverser, state-of-the-art stepper motor drivers, pressure transducers, and programing approach. The new traverser replaces the old belt driven traverser with a mechanical system operated by a few linear translation stages. The new traversing system allows for effective adaptive gridding with much higher spatial resolution and position accuracy when compared to the belt driven system. The new system has increased the number of data points that can be collected in a two-hour turbine run period from 336 points to 868 points, an increase of approximately 160%. Current system parameters are monitored in real-time using a graphical user interface allowing for ease of tracking and monitoring the results of the test.

The current adaptive grid definition is proved to the computer by the user in a manual manner. However, the present system also allows a fully automatic/adaptive grid definition starting from a coarse grid measurement. The present system has capabilities to detect large and small gradients of measured quantities in a relatively coarse grid. The computerized approach can define the new time optimized adaptive grid without human intervention. Experiments of this fully automated measurement grid generation system are underway.

The current approach provides a much needed mass averaged flow measurement system since it can provide all three components of the velocity vector over an area of interest with improved accuracy.

The calibrated probe currently maps the flow field behind the NGV using the new traversing system along one vane pitch and full span. Typical NGV passage exit flow structures such as vane wake, boundary layer, and passage vortex within the field are identified. The adaptive gridding method allows for measurements locations in these flow structures to be increased, while areas of smaller gradients are coarser. This leads to a reduction in the amount of time required to measure one entire vane passage.

The present paper presents significant improvements in FHP based aerodynamic measurements in four significant areas. The specific approach reduces the elapsed calibration time of a typical Five-Hole Probe. A second major improvement is in the spatial resolution of measurements in selected high gradient areas. The third important property of the present approach is in the improved accuracy of the measurements. Finally, the current approach reduces turbine facility run-time significantly.

ACKNOWLEDGMENTS

The authors acknowledge the financial support provided for Jason Town by the Department of Aerospace Engineering at Penn State University. Cengiz Camci also acknowledges the support generously provided to him by TUBITAK, The Scientific and Technological Research Council of Turkey, during his sabbatical leave at Istanbul Technical University. Mr. Rick Auhl, Mark Catalano, and Kirk Hellen of Aerospace Engineering at Penn State provided significant technical expertise during the construction and execution of the experiments. The Five-Hole Probe construction at the subminiature scale would not have been possible without Mr. Harry Houtz's technical contributions.

REFERENCES

1.   B. G. Wiedner, Passage Flow Structure and its Influence on Endwall Heat Transfer in a 90 Turning Duct, The Pennsylvania State University, University Park, Pa, USA, 1994.

2.   L. Treaster and A. M. Yocum, "The calibration and application of five-hole probes," ISA Transactions, vol. 18, no. 3, pp. 23–34, 1979

3.   Ostowari and W. H. Wentz, "Modified calibration techniques of a five-hole probe for high flow angles," Experiments in Fluids, vol. 3, no. 1, pp. 166–168, 1983.

4.   F. R. Nowack, "Improved calibration method for a five-hole spherical Pitot probe," Journal of Physics E: Scientific Instruments, vol. 3, no. 1, pp. 21–26, 1970. ·

5.   J. P. Weiz, An Algorithm for Using the Five-Hole Probe in the Non-Nulled Mode, The Applied Research Laboratory of the Pennsylvania State University, University Park, Pa , USA, 1980.

6. B. A. Reichert and B. J. Wendt, A New Algorithm for Five-Hole Probe Calibration and Data Reduction and Its Application to a Rake-Type Probe, ASME-FED, 1993.

7. R. G. Dominy and H. P. Hodson, "Investigation of factors influencing the calibration of five-hole probes for three-dimensional flow measurements," Journal of Turbomachinery, vol. 115, no. 3, pp. 513–519, 1993.

8. G. L. Morrison, M. T. Schobeiri, and K. R. Pappu, "Five-hole pressure probe analysis technique," Flow Measurement and Instrumentation, vol. 9, no. 3, pp. 153–158, 1998.

9. S. W. Lee and T. J. Yoon, "An investigation of wall-proximity effect using a typical large-scale five-hole probe," Korean Society of Mechanical Engineers International Journal, vol. 3, no. 13, pp. 273–285, 1999.

10. M. C. Brophy, A. L. Treaster, D. R. Stinebring, and J. P. Weiz, "Optimization of a five-hole probe wake measuring system," Technical Memorandum 87-156, Applied Research Laboratory, The Pennsylvania State University, University Park, Pa, USA, 1980.

11. N. Sitaram, B. Lakshminarayana, and A. Ravindranath, "Convential probes for the relative flow measurement in a rotor blade passage," in Proceedings of the Joint Fluids Engineering Gas Turbine Conference and Products Show, New Orleans, La, USA, March 1980.

12. J. Town and C. Camci, "Sub-miniature five-hole probe calibration using a time efficient pitch and yaw mechanism and accuracy improvments," in Proceedings of the ASME International Gas Turbine Conference, ASME GT-46391, Vancouver, Canada, 2011.

13. J. Pisasale and N. A. Ahmed, "Theoretical calibration for highly three-dimensional low-speed flows of a five-hole probe," Measurement Science and Technology, vol. 13, no. 7, pp. 1100–1107, 2002.

14. J. Pisasale and N. A. Ahmed, "A novel method for extending the calibration range of five-hole probe for highly three-dimensional flows," Flow Measurement and Instrumentation, vol. 13, no. 1-2, pp. 23–30, 2002.

15. J. Pisasale and N. A. Ahmed, "Development of a functional relationship between port pressures and flow properties for the calibration and application of multihole probes to highly three-dimensional flows," Experiments in Fluids, vol. 36, no. 3, pp. 422–436, 2004.

16. S. J. Lien and N. A. Ahmed, "An examination of suitability of multi-hole pressure probe technique for skin friction measurement in turbulent flow," Flow Measurement and Instrumentation, vol. 22, no. 3, pp. 215–224, 2011.

17. K. Kuisoon, B. G. Wiedner, and C. Camci, "Turbulent flow and endwall heat transfer analysis in a 90° turning duct and comparisons with measured data—part II: influence of secondary flow, vorticity, turbulent kinetic energy, and thermal boundary conditions on endwall heat transfer," International Journal of Rotating Machinery, vol. 8, no. 2, pp. 125–140, 2002.

18. K. Kuisoon, B. G. Wiedner, and C. Camci, "Turbulent flow and endwall heat transfer analysis in a 90° turning duct and comparisons with measured data. Part I. Influence of Reynolds number and streamline curvature on viscous flow development," International Journal of Rotating Machinery, vol. 8, no. 2, pp. 109–123, 2002.

19. Camci and D. H. Rizzo, "Secondary flow and forced convection heat transfer near endwall boundary layer fences in a 90° turning duct," International Journal of Heat and Mass Transfer, vol. 45, no. 4, pp. 831–843, 2001.

20. Lakshminarayana, C. Camci, I. Halliwell, and M. Zaccaria, "Design and development of a turbine research facility to study rotor-stator interaction effects," International Journal of Turbo & Jet Engines, vol. 13, no. 3, pp. 155–172, 1996.

21. J. R. Taylor, An Introduction to Error Analysis the Study of Uncertainties and Physical Measurements, University Science Books, Sausalito, Calif, USA, 1997.

Chapter 10

DETECTION OF ROTOR FORCED RESPONSE VIBRATIONS USING STATIONARY PRESSURE TRANSDUCERS IN A MULTISTAGE AXIAL COMPRESSOR

William L. Murray III and Nicole L. Key

Purdue University, 500 Allison Road, West Lafayette, IN 47907, USA

ABSTRACT

Blade row interactions in turbomachinery can lead to blade vibrations and even high cycle fatigue. Forced response conditions occur when a forcing function (such as impingement of stator wakes) occurs at a frequency that matches the natural frequency of a blade. The objective of this research is to develop the data processing techniques needed to detect rotor blade vibration in a forced response condition from stationary fast-response pressure transducers to allow for detection of rotor vibration from transient data and lead to techniques for vibration monitoring in gas turbines. This paper marks the first time in the open literature that engine-order resonant response of an embedded bladed disk in a 3-stage intermediate-speed axial compressor was detected using stationary pressure transducers. Experiments were performed in a stage axial research compressor focusing on the embedded rotor of blisk construction. Fourier waterfall graphs from a laser tip timing system were used to detect the vibrations after applying signal processing methods to uncover these pressure waves associated with blade vibration. Individual blade response was investigated using cross covariance to compare blade passage pressure signatures through resonance. Both methods agree with NSMS data that provide a measure of the exact compressor speeds at which individual blades enter resonance.

INTRODUCTION

Blade row interactions in turbomachinery can lead to blade vibrations and even high cycle fatigue. Forced response conditions occur when the frequency of the forcing function matches the frequency of the blade vibration mode. Forcing functions include the viscous wakes shed from upstream blade rows and the potential fields generated by the upstream and downstream blade rows. It is impossible to remove all resonant crossings from the entire operating range of a multistage axial compressor, and thus, it is important to detect blade vibration to ensure it is within proper limits and to bring to light unexpected engine-order resonant conditions. Since recent turbomachinery designs have moved toward higher pressure ratios and integrally bladed rotors (IBRs), aerodynamic forcing environments have grown in strength and mechanical system damping has decreased. It is now more important than ever to develop robust, dependable methods to detect, characterize, and help mitigate potential vibrational issues before they can have disastrous effects.

Common techniques for detecting blade vibration include strain gauges or tip timing probes. Strain gauge techniques can be employed in a number of ways, but the most common technique involves mounting strain gauges on the blades, and when rotor vibrations are considered, a slip-ring is needed to transmit the signals from the strain gauges to the recording equipment. This measurement is easier if, instead, the strain gauges are mounted to stationary hardware, such as a stator vane or used in a cascade experiment, as done by Freund et al. [1]. Tip timing systems, or nonintrusive stress measurement systems, (NSMS) utilize a set (typically 8) of light probes that are arranged in an optimal configuration to detect a particular vibrational mode. Thus, they are limited in the quantity modes that can be measured, and if there is little deflection in the outer part of the blade for the mode of interest, they will not provide any useful information.

Some researchers have also used pressure transducers to study blade vibrations. Pressure transducers can be mounted either on a stationary part of the machine, most often the casing endwall as in the work conducted by Baumgartner et al. [2], or they can be mounted onto the rotating hardware, for example Gill and Capece [3], where the signal would be transmitted through a slip ring device to a recorder. When measuring rotor vibrations in the stationary reference frame, the vibration frequency will be Doppler shifted as discussed by Mengle [4] and Kurkov [5].

The acoustic environment inside of an operating compressor is noisy and usually dominated by flow physics unrelated to compressor vibration. Baumgartner et al. [2] used a single hot film downstream of the tip of a vibrating rotor blade and some casing-mounted pressure transducers to investigate rotating stall and flutter vibration of rotating blades. They found that the spectral magnitudes of the instability and vibration were significantly smaller in amplitude than the spectral magnitude related to blade passing frequency. Kurkov [5] showed that the spectrum of the signal obtained when the blade was not vibrating could be subtracted from the spectrum of the pressure signal measured during the forced response vibration to view the frequency spectrum of forced response. This method has also been employed by Mengle [4] in his attempt to remove integral engine order frequencies from observed spectra of a rotating compressor blade. Kurkov [5] was able to take an average frequency response of 16 revolutions of data without vibration and subtract that frequency spectrum from the vibratory response spectrum.

Rotating stall and flutter have been studied extensively in the realm of aeromechanics. These non-EO vibrations tend to have larger vibration amplitudes and correspondingly larger pressure wave amplitudes, making them easier to measure with pressure transducers. Leichtfuss et al. [6] measured non-engine-order stall flutter spectral signatures with stationary Kulite pressure transducers. Schoenenborn and Breuer [7] measured torsional vibration and flutter of blades during surge and were able to detect aeroacoustic signatures from unsteady pressure transducers related to blade motion relatively easily. They were also able to relate the pressure measurements to NSMS tip timing blade deflection data.

The development of techniques that allow pressure transducers to detect forced response during compressor operation as a real-time diagnostic tool or simply as a backup to more complex vibration detection instrumentation is valuable. Pressure fluctuations due to engine order (EO) vibrations such as forced response are typically smaller in comparison to pressure fluctuations due to non-EO vibrations (such as those associated with flutter, rotating stall, and buffeting), making forced response detection difficult. Additionally, forced response pressure fluctuations are Doppler shifted to blade pass frequencies (and their harmonics), which are already heavily influenced by aerodynamics even when the blades are not vibrating, making the identification of forced response pressure components difficult.

The authors are not aware of any open litureature that discussses directly measured engine-order forced response through the use of stationary pressure transducers in an experimental compressor facility, as the amplitudes of the acoustic waves are generally very small in comparison to other flow field features. Most research conducted to this point has been focused on flow measurements associated with nonintegral engine order vibrations, such as stall, flutter, surge, or acoustic resonances. Fridh et al. [8] measured the spectral component of forced response from partial admission inlet distortion in a turbine using pressure transducers in a rotating reference frame. They were able to detect resonance at Campbell diagram crossings with strain gauge and pressure transducer data. However, the strain gauge data provided more detailed spectral data on the response characteristics of important Campbell crossings compared to the pressure data. Therefore, the objective of this paper is to show how embedded rotor vibrations associated with forced response have been detected with stationary Kulite pressure transducers and describe in detail the important aspects of the instrumentation and data processing techniques that made this possible. This paper marks the first time in the open literature that engine-order resonant response has been measured in a compressor using stationary pressure transducers.

MATERIALS AND METHODS

The Purdue Three-Stage Research Compressor is a unique research facility that models the rear stages of a high-speed compressor, matching Mach number and Reynolds number to aircraft engine operating conditions. The facility is conducive to detailed flow measurements in the pitch wise direction because each vane row can be individually indexed past stationary probes. There is a significant and measurable total pressure rise per stage, and blade heights are 50.8 mm, allowing sufficient space for detailed flow measurements without probe blockage issues.

The facility layout is shown in Figure 1. Unconditioned ambient air is drawn into a large settling chamber. Air enters the inlet duct through a bellmouth, which is followed by a series of flow straighteners. An ASME-standard long-form Venturi flow meter installed in the inlet ducting measures the mass flow rate through the compressor. Following an additional length of insulated ducting inside the test cell, a nosecone directs the flow into the annulus of the compressor. The compressor has a constant-area annulus with a 609.6 mm outer diameter. After passing through the compressor, the air encounters a sliding-annulus throttle and exhausts to atmosphere through a collector. The compressor is driven by a 1 MW AC motor with a variable frequency drive. The motor is connected to a speed-increasing gearbox via a gear coupling to provide the compressor design speed of 5,000 RPM.

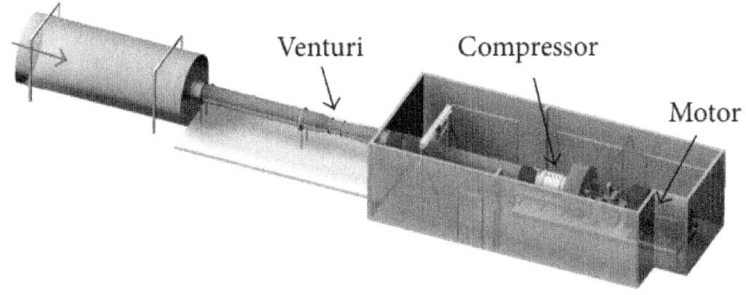

Figure 1: Layout of the Purdue 3-Stage Axial Compressor Research Facility.

The compressor consists of an inlet guide vane (IGV) row followed by three stages, Figure 2. The compressor features IBRs with rotor counts of 36 for Rotor 1, 33 for Rotor 2, and 30 for Rotor 3. The rotors are attached to a drum in a fixed configuration. The shrouded stator rows have similar vane counts of 44 except for S3, which has 50 vanes. The vane rings are split into two halves and installed in the split casing. Steady compressor performance is measured with 7-element Kiel head total pressure and total temperature rakes. The overall compressor total pressure ratio is measured with rakes positioned at Stations 0 and 9.

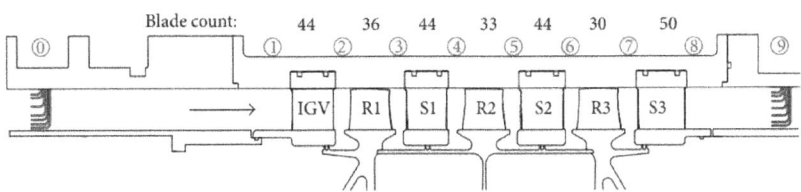

Figure 2: A cross-section of the flow path including data acquisition stations.

Figure 3 shows the Campbell diagram for Rotor 2. With 44 vanes upstream in Stator 1 and 44 vanes downstream in Stator 2, the 44 engine order excitation of the first torsion (1T) vibratory mode is the Campbell diagram crossing studied in this research. The frequency of the 1T mode is 2700 Hz, and the 44EO excitation of the R2 1T vibratory mode occurs near 3700 RPM (74% speed). Also shown is the first torsion mode shape and nodal line as calculated by Fulayter [9] using a finite element analysis.

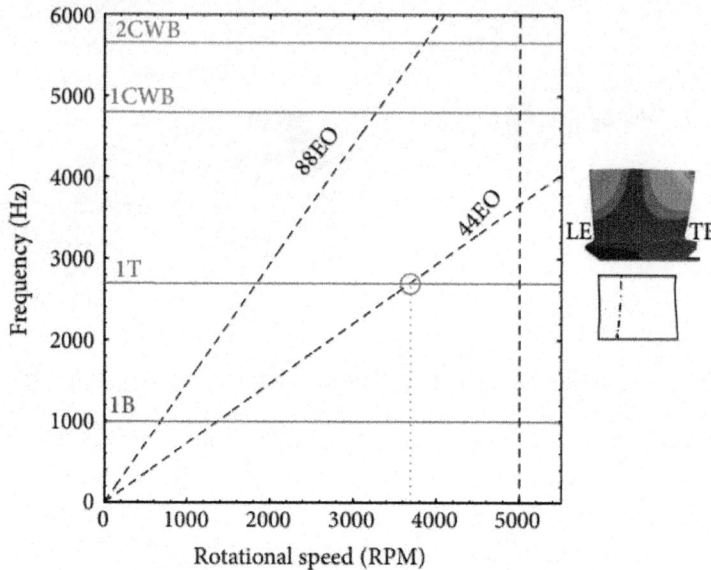

Figure 3: Campbell diagram for Rotor 2 including mode shape and nodal line for the 1T vibratory mode.

Tip timing data were acquired to characterize the Rotor 2 1T vibratory response to the 44EO excitation by measuring the rotor tip deflection. These measurements can be used to compare the results obtained with the new data processing technique developed for the pressure sensors. The Agilis nonintrusive stress measurement system (NSMS) consists of 8 fiber-based optical probes, laser and detector boxes, and an NI 5112 ADC data acquisition chassis. The laser module generates a signal that is sent to each of the 8 probes that shine down on the passing blades. The casing-mounted probes also have a sensing optical cable that transmits the reflection of the blade tips to a photo detector and a pulse-to-digital converter. The converter has a digital clock to track the time of the blade passing and relate it to a 1/rev shaft signal. The timing of each blade arrival and the correlation of arrival times over all 8 sensors allow the measurement of the amplitude and phase of vibration for each blade.

The resolution of the tip timing measurements is a function of the counter timer board clock speed, the rotor tip diameter, and the rotor tip speed. With a sampling period of 2 microseconds, the resolution of the tip timing data near the 1T vibratory resonance of Rotor 2 is about 0.009 mils. Since it is unsafe to operate at the resonance speed due to potential fatigue and failure of the vibrating rotor, the data are acquired during a transient speed change through the resonant condition, and the sweep rate is an important parameter. As suggested by von Flotow [10], the critical sweep rate is defined as the ratio of the half-power bandwidth acquired over 3 vibration cycles. For the 1T resonance of Rotor 2, the critical sweep rate is around 15 rpm/s. Also, the control of the compressor drive system was programmed such that the compressor speed would have a constant acceleration rate through the resonant condition, and thus, this aspect of the experiment was fully controlled. Sweep rates as low as 2.2 RPM/s were utilized.

The steady loading of the compressor was adjusted using the throttle. Two loading conditions were considered: nominal loading (NL) and high loading (HL). Figure 4 shows the compressor performance map including two speedlines: 68% and 80% corrected speed. These speedlines bound the resonant condition studied. Each point on the speedline represents the area-averaged pressure ratio as determined from a vane traverse every 5% passage to include the effects of the vane wakes. The mass flow rate has been normalized by the stall flow rate at 68% corrected speed. The total pressure ratio is the overall compressor total pressure ratio. The operating lines were created by quickly scanning the pressure rakes and flow meter while accelerating the compressor through resonant speed, and thus, they are not area-averaged quantities.

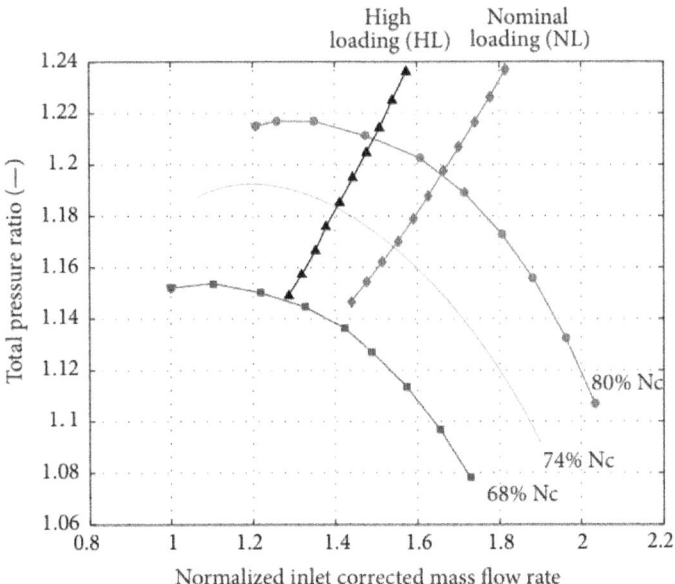

Figure 4: Compressor operating map including operating lines.

The pressure measurements were acquired from the downstream vane which was instrumented for a different research project. The sensors were Kulite LQ-062 pressure transducers with a range of 0–5 psig. The key aspect that made these sensors useful in blade vibration detection is that they had no screen, and this resulted in a high frequency response, as much as 100–150 kHz per manufacturer's specifications. Other sensors in the facility with lower frequency response did not capture the same phenomenon. To briefly describe the installation of these sensors, a removable S2 passage that could be sent to Kulite for sensor installation was fabricated, Figure 5. The part was constructed of 17-4 PH stainless steel and EDM cut into 3 segments. The vanes of the removable passage were machined to accommodate a total of 16 transducers, 8 on each vane. They were positioned at 50% span and 80% span, and axial positions include 10%, 20%, 30%, and 40% chord on the pressure side of one blade and the suction side of another. The sensors were clustered as close as possible to the leading edge, and RTV was used to back fill the pockets drilled for the sensors.

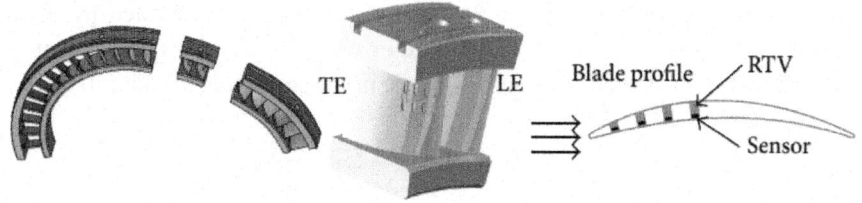

Figure 5: Vane ring modification for accommodation of Kulite sensors.

Figure 6 shows a diagram of the location of the pressure transducers installed in Stator 2 with respect to the flowpath. Because of space restrictions, a single vane could not support transducers on both the pressure side and suction side. Thus, a passage was instrumented, with the transducers measuring a particular flow passage. This arrangement was also favored because RTV was used to backfill the pocket, and thus, the RTV did not affect the pressure measurements since measurements were not made on the same surface with the RTV treatment.

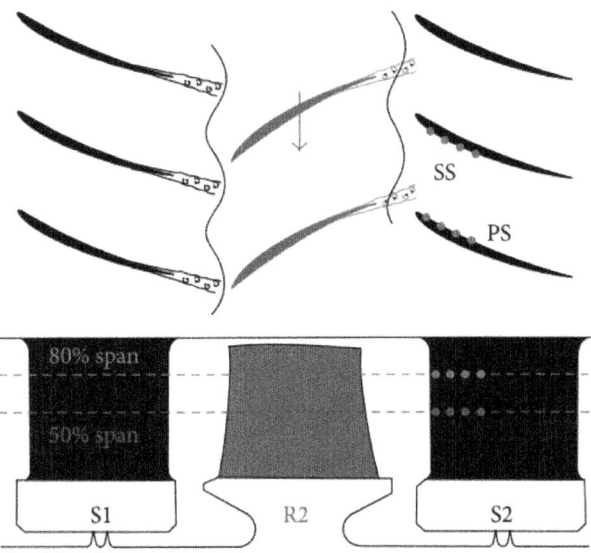

Figure 6: Top-down and side view of Stator 2 with installed pressure transducers.

Figure 7 shows photographs of the removable Stator 2 insert with the Kulite pressure transducers installed. The wires are fed through a hole in the casing and the vane row is not traversed when this passage is installed. In the photograph on the right, the RTV on the suction side of the top vane is visible. The unsteady pressure on the pressure side of the top vane is measured. Also, some of the holes for the pressure measurements on the suction side of the bottom vane are visible.

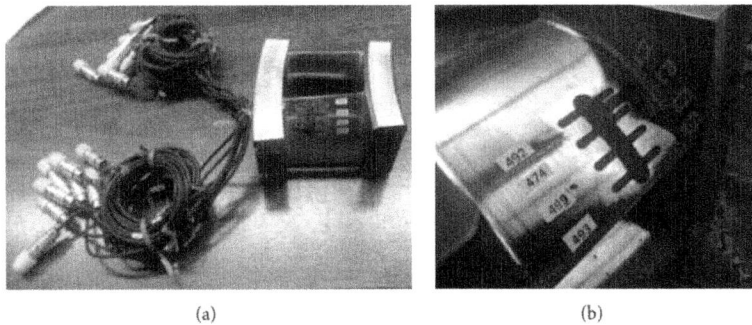

(a)   (b)

Figure 7: Photographs of the removable Stator 2 insert with pressure transducers installed.

The excitation signal and amplification for the Kulites were provided by a Precision Filter 28000 chassis with quantity 4, 28118 full bridge amplification cards, each capable of amplifying 8 channels. Data were digitized with an NI PXI-1073 chassis with two 16-channel NI PXIE-6358 cards, with a total bandwidth of 1.25 MS/s per channel for all 32 channels simultaneously sampled. The transducers were calibrated with a calibration chamber, Figure 8, constructed from 15.24 cm cast aluminum pipe that was 15.24 cm long with caps on each end. The stator was installed in the chamber by a short piece of threaded rod that was threaded into one of the machined and tapped threads in the shroud of the stator. The LEMO connectors were fed out of two holes drilled in one of the end caps and sealed with silicone sealant. The heat shrink used on the outside of each of the wires was also sealed at each end with silicone to keep air from leaking through the heat shrink and out of the calibration chamber. The slope of the calibration for the installed stator sensors was as much as 25% different from the values quoted by Kulite for the sensors prior to installation in the stator highlighting the need for calibration.

(a)   (b)

Figure 8: Calibration chamber setup for the vane-mounted pressure transducers.

RESULTS AND DISCUSSION

Since this is a research compressor that operates at lower pressure levels than an actual gas turbine compressor, the pressure waves associated with blade vibration are expected to be small. To assess the ability of the available pressure instrumentation to detect the blade vibration, a simple analysis was performed. LINSUB, a linear, flat plate cascade, aeromechanics solver created by Whitehead [11], was used to predict the expected pressure amplitude due to the vibration of Rotor 2 for this forced response condition. It outputs the absolute and fluctuating components of the flow field (such as lift, pressure, and moments) related to bending, torsion, chord wise bending, and wakes/gusts. In this case, it is of interest to calculate the unsteady pressure waves traveling upstream and downstream from a blade vibrating in torsion. Based on the compressor geometry and previously measured flow conditions near resonance, LINSUB predicted an upstream-traveling pressure component of 292.3 Pa and a downstream-propagating pressure wave of amplitude 584.7 Pa.

The typical uncertainty of the Kulite pressure transducers model LQ-062 is ±0.1% of the full range of the transducer, which corresponds to ±34.5 Pa. (The manufacturer provides a typical and maximum uncertainty, where the maximum is ±0.5%.) This uncertainty does not include any uncertainty or noise introduced by connectors, cables, Precision Filter amplifier equipment, or the PXI analog-to-digital conversion. By including these effects, the largest resulting uncertainty in the Kulite pressure measurements amounts to approximately ±175.8 Pa. This was calculated considering that the Precision Filter 28118 amplifier cards have an accuracy of ±0.1% of full scale range and the PXI-6358/6356 A/D converters have an uncertainty of 0.012% full scale.

The LINSUB analysis is 1D, and thus serves as an approximation of the expected pressure amplitudes due to rotor vibration. Nonetheless, the results indicate that at these compressor speeds, the pressure waves due to resonance are small. However, if signal processing techniques can be developed to identify the blade vibration effects at these low compressor speeds, then vibrations in actual compressors which operate at higher speeds should be easier to detect.

To detect these small pressure waves generated by the vibrations of Rotor 2, several data processing methods were investigated. The most difficult issue was the removal of background noise and pressure information that was not pertinent to the blade vibration. Additionally, data handling algorithms to manage the large volumes of unsteady binary data were developed.

The Kulite pressure transducers are in the stationary reference frame, but the vibrating rotor is in the relative reference frame, and thus, the frequency of vibration as measured by the transducer will be shifted due to the relative motion between the rotor and the transducer. Mengle [4] showed that since the rotor is spinning at a rotation rate, Ω, the vibration frequency of the blades, ω, will be Doppler shifted to a different frequency, ω', based on nodal diameter, ND, and wave number, m, in (1).

$$\omega' = \omega + (\text{ND} + mB)\,\Omega.$$

(1)

The nodal diameter can either be determined from the NSMS data or from the difference between stator vanes and rotor blades, since the frequency of vibration is an engine-order (EO) forced response, in this case, as shown in

$$ND = N_{\text{Stator Vanes}} - N_{\text{Rotor Blades}} = 44 - 33 = 11. \tag{2}$$

Because of blisk mistuning, the approximate band of blade response varies over a range of 2700–2735 Hz, as measured with NSMS. By choosing m values of −2, −1, 0, 1, and 2, the first few wave modes will be captured. These should be the highest responding modes. When the frequency of interest is associated with an engine-order forced response vibration, the Doppler shift will always shift the forced response to multiples of the blade pass frequency. Therefore, the signal processing challenge is to determine changes in the 33/rev frequency (and higher harmonics) in the spectrum. This may prove difficult because as the compressor rotational speed increases, changes in the strength of these components of the signal will also be associated with aerodynamic changes. However, if a spike in the higher harmonics of the signal is visible only in a RPM band where rotor vibration is known to occur (as measured with the tip timing data), then it will confirm that the pressure sensors detected rotor vibration.

An effective way to analyze the change in the pressure spectrum with time is with Fourier transform waterfall graphs. A cartoon explaining the construction of these graphs is shown in Figure 9. The pressure data from the sensors are recorded as the compressor rotational speed is increased at a constant rate through the speed corresponding to the first torsion vibratory mode. First, the pressure data from a particular channel are divided up into separate compressor revolutions using the use of the once-per-revolution signal from the compressor shaft. Simply dividing the data by each revolution is effectively square windowing the signal, which can introduce unwanted artifacts into the frequency spectrum, and thus, a Kaiser window was applied to each revolution. A Kaiser window allows a high resolution between two frequency components that are separated greatly in amplitude but closely related by frequency. This windowing technique allows for the detection of small components of the 1T signal that are Doppler shifted to a frequency slightly different than blade pass frequency due to individual blade mistuning. It also allows for a cleaner waterfall plot, resolving features due to a better resolution in frequency.

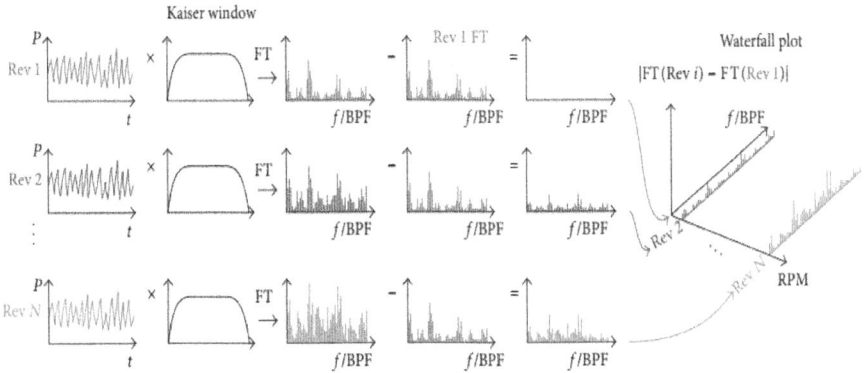

Figure 9: Data processing strategy for generating waterfall graphs.

After windowing the revolution of data, a Fourier transform is performed, and the frequency is normalized by the rotor blade pass frequency. This normalization is important because it connects the spectral magnitudes to blade pass events rather than to physical compressor speeds, which are constantly changing through the sweep. Pressure fluctuations will occur with the blade passing events due to rotor wakes impinging on the downstream vane row. Thus, a significant magnitude in the spectral component associated with blade pass frequency will always be present. To reduce this effect in the hopes of identifying the small pressure waves associated with the resonant vibration, the spectral magnitudes measured at the beginning of the sweep (3670 rpm) are subtracted from the spectra at all successive revolutions. Then, these reduced spectra are assembled to create the resulting waterfall graph. This provides a snapshot of several thousand Fourier transforms, allowing trends to be easily identified. This also allows for easy comparison of the changes in amplitude of a given frequency band as the compressor accelerates through resonance.

Figure 10 shows the waterfall graph constructed from data acquired at nominal loading. The labels on the right show particular blade pass frequencies and their harmonics. Recall the blade counts decrease by 3 (Rotor 1 has 36 blades, Rotor 2 has 33, and Rotor 30 has 30). The 44EO excitation of the Rotor 2 first torsion vibration should occur at Rotor 2 blade pass frequency (33/rev) or its harmonics. The 33/rev frequency component is, for the most part, nonexistent. Also, the 66/rev tends come and go. However, the 99/rev either is nonexistent or appears and disappears in conjunction with resonance. There are also other frequencies present, namely, responses at 30/rev, 36/rev, higher order harmonics of these frequencies (which correspond to R1 and R3 blade pass frequencies). This is due to the pressure transducers picking up the destructive and constructive interference of R1 and R3. The manner in which the reflected multiples of these blades pass frequency harmonics is transmitted through the compressor and could be indicative of resonance, perhaps even R3 resonance which occurs in the speed range of 3705–3805 RPM.

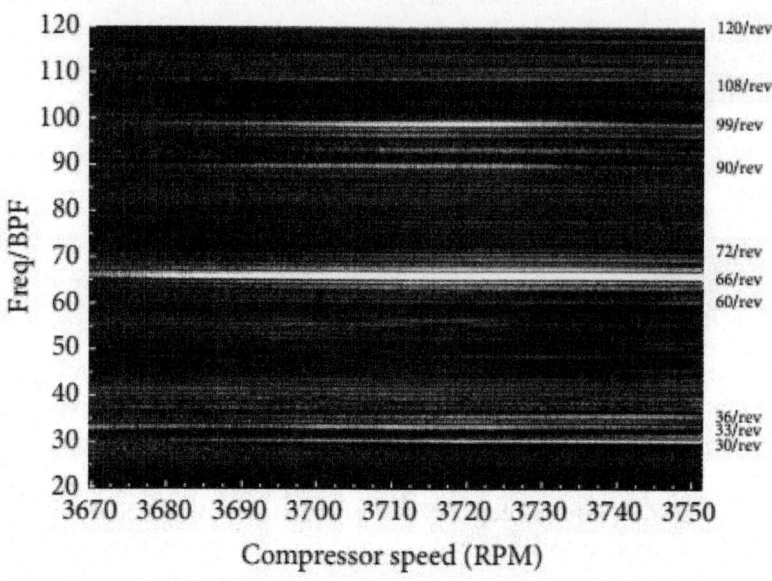

Figure 10: Waterfall graph at nominal loading.

Figure 11 shows the amplitude of the 99/rev response (taken to be the sum of the amplitudes in the frequency band between 98.5/rev and 99.5/rev), and the response grows and decays as a function of compressor speed. For comparison to NSMS data, Figure 12 shows the RPM band of the Rotor 2 resonant response. Each blade has its own natural frequency, and the results from all 33 blades are overlaid. One graph shows the results at nominal loading and the other at high loading. The mean loading does not change the frequency of the response, but it does change the amplitude of the response. The compressor speeds (or frequency) of the 1T response ranges from approximately 3700 to 3730 RPM. The range at which the 99/rev frequency bands increase in amplitude from the pressure transducer measurements agrees well with the NSMS data.

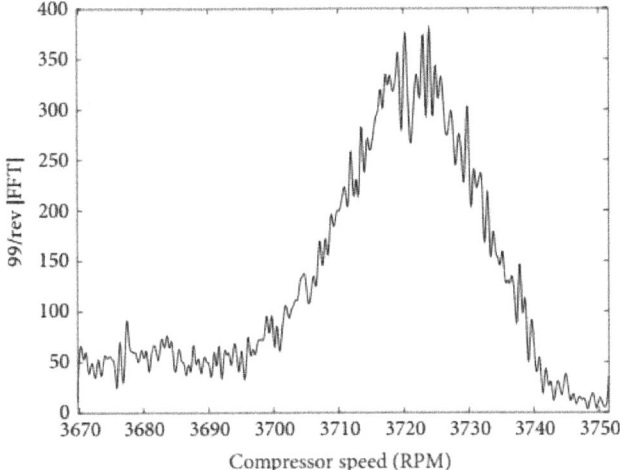

Figure 11: The magnitude of the spectral component associated with the 99/rev frequency at nominal loading.

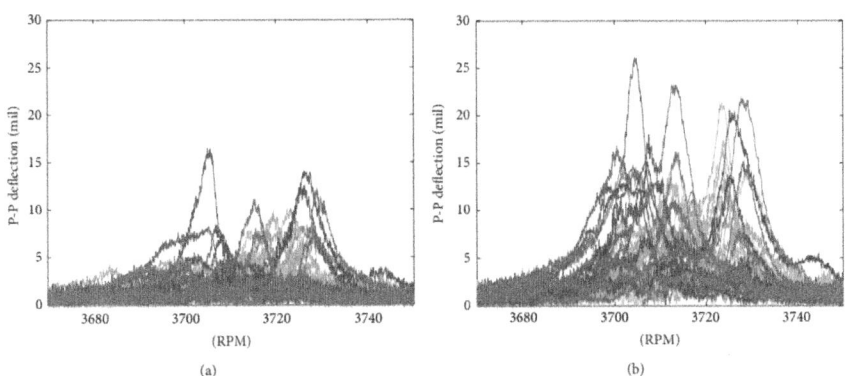

Figure 12: NL (a) and HL (b) peak-to-peak deflection of each blade of R2 as recorded by NSMS.

The 66/rev response grows but then seems to level out as RPM is increased, and this could be attributed to the data processing technique, where the nonvibrating spectrum at the low speed range of the sweep was used to adjust the other spectra, as shown in Figure 9. Therefore, if the spectral magnitudes change because of the change in compressor speed resulting in a change of the aerodynamic interaction of the blade rows, then that would appear in the waterfall graphs as well.

The 33/rev spectral component does not have a significant amplitude, and this is a result of the data processing methods. The blade pass frequency component is large and by subtracting the spectrum from the first revolution off of all subsequent signals, it was sufficient to remove the blade pass component from the other revolutions. The spectral magnitude of blade pass frequency was an order of magnitude higher than the harmonics, and thus, it would be difficult to identify the small pressure fluctuations associated with rotor vibration at this frequency. Therefore, to detect forced response vibration effects from the vane pressure signal, efforts should focus on the higher harmonics of blade pass frequency where the Doppler shifted forced response frequency components are on the same order of magnitude as the amplitudes that occur when the blade is not vibrating.

The waterfall graph for the results at high loading is shown in Figure 13. As before, the magnitude of the 99/rev spectral component signal is shown in Figure 14. The increased spectral magnitude of the 99/rev frequency occurs over the same speed range as the NSMS system measured the rotor vibration, as shown in Figure 12. The 33/rev and 66/rev frequencies in the waterfall graph follow the same trends as shown at nominal loading: the 33/rev contribution has been subtracted out and the 66/rev contribution seems to be increasing with compressor speed.

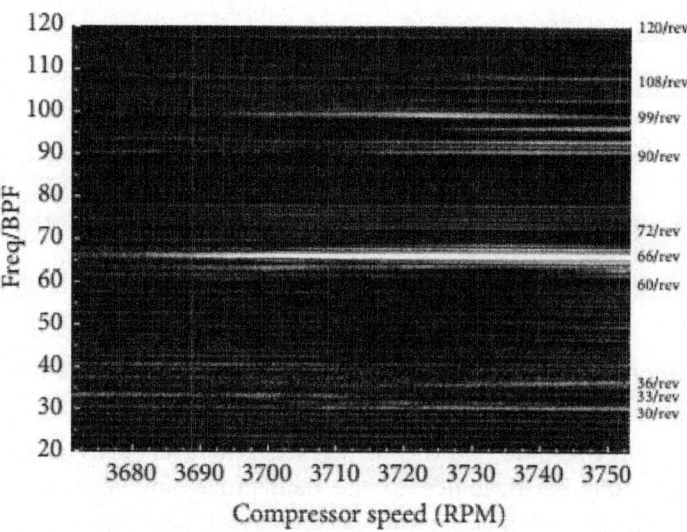

Figure 13: Waterfall graph at high loading.

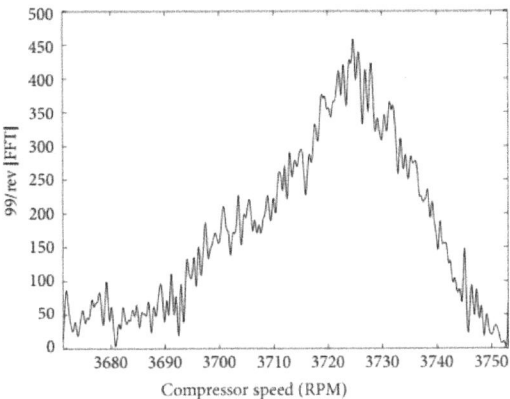

Figure 14: The magnitude of the spectral component associated with the 99/rev frequency at high loading.

In summary, the frequency domain analysis showed that the pressure transducers did detect the rotor vibration at a Doppler shifted frequency associated with 3rd harmonic of blade pass frequency (99/rev). This waterfall analysis considers the spectrum on a revolution-by-revolution basis. Another way to investigate these data is to focus on a time domain analysis where individual blade pass events are considered, rather than a full revolution of data. The advantage of having NSMS data is the vibration frequency and amplitude of each rotor blade is recorded, and thus, the pressure signals associated with highly responding blades can be investigated. This can be done with a cross correlation which essentially looks for a blade pass event that appears to be different than the others. The pressure trace measured when the blade is vibrating will be different than that measured when the blade is not vibrating, and thus a low correlation value could be an indicator of the forced response event.

Cross correlation and cross covariance are methods of determining the "similarity" or likeness, of two random vectors or signals, denoted by x and y with length N. By varying the offset, m, between the two vectors, a correlation sequence of the two vectors is generated and is a function of the element-by-element product of each point in the vector, as shown in

$$\hat{R}_{xy}(m) = \begin{cases} \sum_{n=0}^{N-m-1} x_{n+m} y_n^*, & m \geq 0 \\ \hat{R}_{yx}^*(-m) & m < 0. \end{cases}$$

(3)

Cross correlations, however, are not normalized, and since the amplitude of the mean level of pressure continually increases due to the increase in compressor total pressure ratio with increasing speed during the sweep, the transient nature of this experiment renders the cross correlation inappropriate. Therefore, to remove the effects of the magnitude differences, the cross covariance is considered. It is the product of two vectors, which are normalized by their mean, denoted by $(1/N)\sum_{i=0}^{N-1} x_i$ and $(1/N)\sum_{i=0}^{N-1} y_i^*$. The cross covariance, (4), is used since there will be no overall rise in correlation due to the rise in overall compressor pressure ratio. Consider

$$c_{xy}(m)$$

$$= \begin{cases} \sum_{n=0}^{N-|m|-1} \left(x(n+m) - \left(\frac{1}{N}\right) \sum_{i=0}^{N-1} x_i \right) \\ \qquad \times \left(y_n^* - \left(\frac{1}{N}\right) \sum_{i=0}^{N-1} y_i^* \right) & m \geq 0, \\ c_{yx}^*(-m) & m < 0. \end{cases}$$

$$(4)$$

Figure 15 shows the pressure traces measured on the pressure side of Stator 2 when Blade 26 was passing in front of it. The different traces are shown for different compressor speeds to illustrate how the pressure signal associated with a particular Rotor 2 blade is changing throughout the compressor sweep. The red trace at 3671 RPM is the pressure trace from the 1st full revolution of data to which all successive pressure traces are compared for this analysis. The blue and cyan lines, at 3705 and 3710 RPM, respectively, show pressure traces of the blade in and around resonance, and the 3735 RPM black lines shows the pressure trace near the end of the sweep. This illustrates how much the shape of the pressure signal will change as the compressor pressure ratio increases.

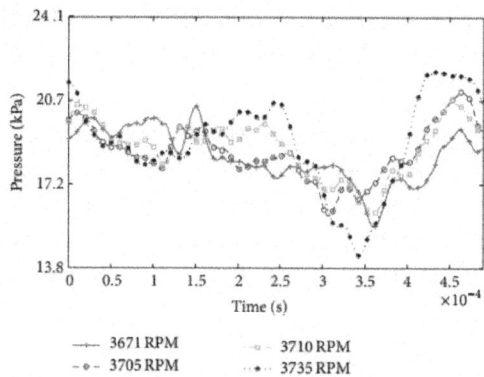

Figure 15: Rotor blade passage traces, Blade 26, PS 80% Span 20% Chord, HL.

In Figure 16, there is a large drop in cross covariance around 3705 RPM, which corresponds directly to the RPM band of maximum displacement in R2 for blade 26 at high loading (refer to Figure 12). Although the drop in covariance amplitude is small in comparison to the revolution-to-revolution changes in covariance amplitude (i.e., the noise in the signal), it is apparent that there is approximately a 30% drop in average covariance amplitude from 3700 to 3705 RPM, and then a corresponding 30% increase at 3705–3710 RPM.

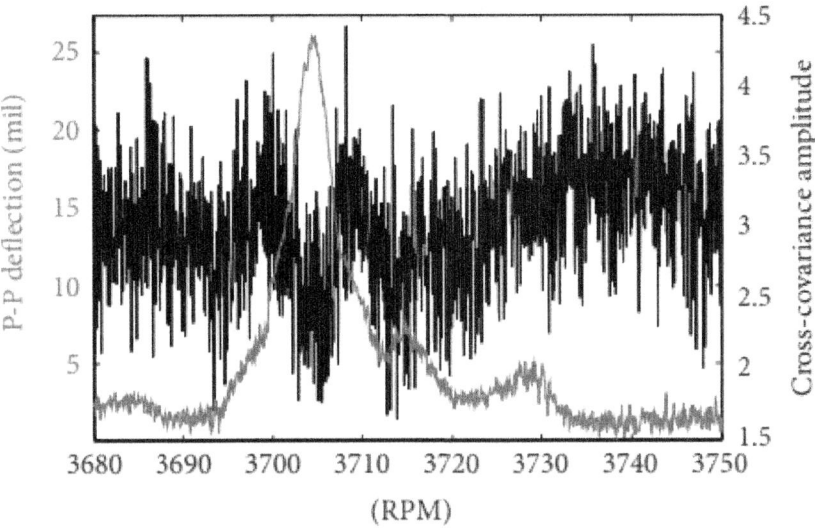

Figure 16: Cross covariance and NSMS deflection for Blade 26 at high loading (80% span).

Blade 22 is another high responding blade according to the NSMS data from Figure 12. The cross covariance of the pressure signal acquired at 80% span shows a similar drop in cross covariance, Figure 17. The compressor speed range over which this occurs is roughly equivalent to the speed range for which the NSMS data show that the blade vibration occurs. There is an inherent waviness, or fluctuation, in the overall trend of the cross covariance that is apparent in all sensors regardless of speed, which makes the measurements of cross covariance amplitude drops difficult to compute. Future work is aimed at advanced signal processing techniques to further explore the utility of the cross covariance in identifying rotor vibration.

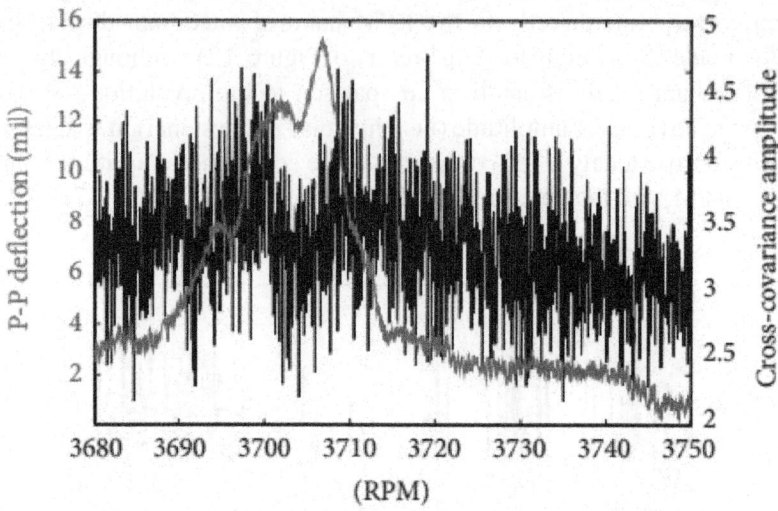

Figure 17: Cross Covariance and NSMS deflection for Blade 22.

CONCLUSIONS

This is the first time in the open literature that measurements of a compressor forced response vibrational mode have been detected with stationary pressure transducers. This paper explains the particular data processing methods used to identify rotor vibration. The data processing was challenging since the data were transient in nature; the compressor was accelerated through the resonant speed as the data were collected to avoid depleting several cycles from the life of the blade. An additional challenge was associated with the fact that forced response rotor vibrations always get Doppler-shifted to a harmonic of blade pass frequency when measured in the absolute reference frame. This makes it difficult to identify the smaller contributions of that spectral magnitude that are associated with vibrations rather than the aerodynamics of the blade passing event. On top of all this, the smaller pressure levels associated with a research compressor result in small-amplitude pressure waves generated when the blade vibrates. Despite all these challenges, the technique was successful.

Detection of engine-order forced response at part speed compressor operation through the use of stationary pressure transducers could have an impact on the way blade vibrations are measured in turbomachinery, as this measurement method is significantly simpler than other advanced vibration measurements, such as rotating instrumentation and/or laser-based tip timing measurements. It can also revolutionize the health monitoring strategy of gas turbine compressors as it can identify high cycle fatigue issues prior to blade failure.

The focus of this research was to interrogate the signals obtained from fast-response pressure transducers while a 3-stage compressor was being accelerated through force response resonant speeds. The 44 engine order excitation of the first torsion vibrational mode of the embedded Rotor 2 was studied. Kulite pressure transducers without screens had enough frequency response to detect the blade vibration. In this experiment, the transducers of this type were installed in the downstream stator vane, but flush-mounting this sensor in the casing over the rotor would be even more convenient. LINSUB was used to estimate whether the Kulite pressure transducers would be able to detect acoustic signatures from this forced response event, and it showed that the acoustic pressure fluctuations would be slightly higher (3-4 times) than the uncertainty in the Kulite pressure measurement. The amplitude of the pressure wave will increase with compressor rotational speed so while the pressure signature associated with blade vibration may be small in this research vehicle, it could be more significant at actual engine operating conditions rendering this technique perhaps even more effective for higher speed compressor investigations.

Data processing in both the frequency and time domain show that the acoustic signature generated by the vibrating rotor is detectable. Through the calculation of Doppler-shifted vibration frequencies, spectra related to the rotor vibration have been detected in the 3rd harmonic of the rotor blade pass frequency. Additionally, through the use of a cross covariance blade-to-blade similitude analysis, individual blade vibration has been detected and confirmed with NSMS tip timing results.

Future work is aimed at advanced signal processing techniques to further explore the utility of the cross covariance in identifying rotor vibration. In particular, the noise in the cross correlation results must be reduced if this type of signal is to be used in a control loop for compressor operation. Also, close collaboration with sensor manufacturers should be aimed at reducing the cost and increasing the durability of these sensors for implementation in actual gas turbines.

ACKNOWLEDGMENTS

This work was funded by the GUIde IV Consortium, which included NASA, the US Air Force, Rolls-Royce, GE, Pratt & Whitney, Honeywell, Siemens, and Mitsubishi. The authors are grateful to the GUIde IV Consortium for its guidance and financial support of this work. Also, the authors acknowledge Natalie Smith's artistic contribution to the figures.

REFERENCES

1. O. Freund, M. Bartelt, M. Mittelbach, M. Montgomery, D. M. Vogt, and J. R. Seume, "Impact of the flow on an acoustic excitation system for Aeroelastic studies," ASME Journal of Turbomachinery, vol. 135, no. 3, Article ID 031033, 2013. · ·

2. M. Baumgartner, F. Kameier, and J. Hourmouzaiadis, "Non-engine order blade vibration in a high pressure compressor," in Proceedings of the ISABE International Symposium on Airbreathing Engines, Melbourne, Australia, September 1995.

3. J. D. Gill and V. R. Capece, "Experimental investigation of flutter in a single stage unshrouded axial-flow fan," in Proceedings of the 42nd AIAA Aerospace Sciences Meeting and Exhibit, pp. 7997–8005, Reno, Nev, USA, January 2004.

4. V. Mengle, "Acoustic spectra and detection of vibrating rotor blades, including row-to-row interference," in Proceedings of the AIAA 13th Aeroacoustics Conference, Tallahassee, Fla, USA, October 1990.

5. P. Kurkov, "Flutter spectral measurements using stationary pressure transducers," Journal of Engineering for Power, vol. 103, no. 2, pp. 461–467, 1981. · ·

6. S. Leichtfuss, F. Holzinger, C. Brandstetter, F. Wartzek, and H. P. Schiffer, "Aeroelastic investigation of a transonic research compressor," in Proceedings of ASME Turbo Expo, San Antonio , Tex, USA, June 2013.

7. H. Schoenenborn and T. Breuer, "Aeroelasticity at reversed flow conditions—part II: application to compressor surge," Journal of Turbomachinery, vol. 134, no. 6, Article ID 061031, 8 pages, 2012. ·

8. J. Fridh, B. Laumert, and T. Fransson, "Forced response in axial turbines under the influence of partial admission," Journal of Turbomachinery, vol. 135, no. 3, Article ID 041014, pp. 041014-1–041014-9, 2013. · ·

9. R. D. Fulayter, An experimental investigation of resonant response of mistuned fan and compressor rotors utilizing NSMS [Ph.D. dissertation], School of Mechanical Engineering, Purdue University, West Lafayette, Ind, USA, 2004

10. A. von Flotow, Hood Technologies Corporation's Analyze Blade Vibration Software Version 3.3 User Manual, 2003.

11. D. S. Whitehead, "Unsteady two-dimensional linearized subsonic flow in cascades," in AGARD Manual on Aeroelasticity in Axial-Flow Turbomachines, vol. 1, pp. 3.24–3.30, 1987.

Chapter 11

EXPERIMENTAL INVESTIGATION OF FACTORS INFLUENCING OPERATING ROTOR TIP CLEARANCE IN MULTISTAGE COMPRESSORS

Reid A. Berdanier and Nicole L. Key

School of Mechanical Engineering, Purdue University, 500 Allison Road, West Lafayette, IN 47907, USA

ABSTRACT

An analysis of compressor rotor tip clearance measurements using capacitance probe instrumentation is discussed for a three-stage axial compressor. Thermal variations and centrifugal effects related to rotational speed changes affect clearance heights relative to the assembled configuration. These two primary contributions to measured changes are discussed both independently and in combination. Emphasis is given to tip clearance changes due to changing loading condition and at several compressor operating speeds. Measurements show a tip clearance change approaching 0.1 mm (0.2% rotor span) when comparing a near-choke operating condition to a near-stall operating condition for the third stage. Additional consideration is given to environmental contributions such as ambient temperature, for which changes in tip clearance height on the order of 0.05 mm (0.1% rotor span) were noted for temperature variations of 15°C. Experimental compressor operating clearances are presented for several temperatures, operating speeds, and loading conditions, and comparisons are drawn between these measured variations and predicted changes under the same conditions.

INTRODUCTION

Rotor tip clearance height is known to be a significant contributor to overall compressor performance. In particular, previous studies have discussed its influence on total pressure rise, efficiency, and stall margin [1]. Current goals of increased efficiency and decreased fuel burn for gas turbine engines are focused toward more aggressive compressor designs featuring increased blade loadings and decreased blade heights in the rear block of high pressure engine cores. In these rear stages, blade aspect ratios are smaller, tip clearance heights are large relative to blade span, and endwall flows contribute to a more significant portion of the overall loss [2]. As a result, research focusing on the underlying flow physics for rotor tip leakage flows is growing in importance, particularly for tip clearance heights which are large as a percentage of overall blade span.

Ongoing research projects typically apply an extensive array of experimental techniques, but the ability to connect these experimental results to numerical models and validate computational tools is essential [3]. Although many computational fluid dynamics models struggle to accurately model the fundamental flow physics of tip clearance flows [4], the overall performance deltas due to clearance changes are typically comparable to experimental results. Small tip clearance changes and the related performance changes which occur as a result of changes to ambient temperature are typically considered negligible [5]. However, recent research has shown the measurability of these small performance changes in a multistage compressor [6]. As a result, it is imperative that the tip clearance heights are known for the conditions at which the experimental performance data are collected so that the compressor system can be appropriately modeled in numerical simulations.

Static rotor tip clearances—sometimes referred to as "cold" clearances—do not typically represent the operating, or "hot" running, clearances. In general, thermal growth, pressure forces, and circumferential forces due to high-speed blade rotation contribute to blade growth which leads to a clearance height change at different operating conditions and for different ambient conditions. Mechanical touch probes and erodible rub sticks offer low-cost options for measuring tip clearance during operation with easy implementation, but these methods only measure the tallest blade (i.e., the smallest clearance) during an entire test campaign [7], and no additional information is gained as a result of changes in operating condition or ambient temperature. Thus, noncontact clearance measurement systems, such as optical probes, eddy current sensors, or capacitive sensors, provide an alternative solution. Optical probes can be prone to failure if their line of sight is blocked by foreign objects, and thus the rugged design of capacitance sensors makes them a primary candidate for gas turbine applications [8].

The introduction of frequency modulated (FM) capacitance probe clearance measurement systems by Chivers [9] and Barranger [3] provided alternatives to earlier optical measurement systems. Continued component improvements have provided the ability to incorporate these high-accuracy systems in increasingly harsh environments. An overview of the advancement of capacitance probe measurements systems is given by Sheard [8].

Many authors have presented methods for modeling variations of tip clearances (e.g., Agarwal et al. [10], Kypuros and Melcher [11], and Dong et al. [12]). However, up to now, greater interest has been given to turbines, instead of compressors, due to the known benefits of active clearance control [13] and more significant thermal effects. Of the limited studies focused on predictive clearance modeling in compressors, Dong et al. discuss clearance changes due to loading condition at a fixed speed. Specifically, the authors noted changes of tip clearance approaching 0.5% span for the rear stages of a 10-stage compressor simulation. Further, simulations yielded misrepresentations of flow rate and efficiency when clearances corresponding to the static, or "cold," assembled configuration were implemented in the solution instead of estimated operating clearances.

While the work of Dong et al. [12] provides an important step in predictive modeling development for compressor tip clearances, there is a lack of experimental data in the open literature to support this model and advance the state of the art. Thus, the work presented here aims to fill that void by experimentally validating the variations of clearance with operating condition that was predicted by Dong et al. Further discussion is given to environmental effects on rotor tip clearances, and comparisons are drawn between the operating clearances measured by a capacitance probe system and the predictive clearances using the model proposed by Dong et al. Several recommendations are given for consideration in future test campaigns since experimental tip clearance studies which also measure the operating tip clearance are largely absent in the open literature.

EXPERIMENTAL METHODS

This research was performed in the three-stage axial compressor research facility at Purdue University. Extensive experimental compressor performance data have been collected for an advanced tip clearance study implementing three different rotor tip clearance heights. In support of this project, a recent facility upgrade has incorporated a capacitance-type tip clearance measurement system to measure the operating tip clearances in the compressor.

Compressor Facility

The Purdue three-stage axial compressor research facility is shown in Figure 1. The compressor draws unconditioned atmospheric air into a large settling chamber. A bell mouth directs the air into a long duct and through an ASME-standard long-form Venturi meter where the mass flow rate is measured. A nosecone directs the air into the 50.8 mm constant-annulus height (with a hub-to-tip ratio of 0.8333 throughout). Past the compressor exit, the air passes through a sliding-annulus throttle and is then directed out a scroll-type collector where the air exhausts to ambient conditions. The compressor is driven from the rear by a 1044 kW AC motor, and the driveline passes through a 5 : 1 speed-increasing gearbox to facilitate a corrected design speed of 5000 rpm. An encoder on the motor drive shaft and a proportional-integral-derivative control sequence maintain the rotational speed of the compressor within 0.01% of the desired set point. Aside from the motor control, an optical laser tachometer aimed at the high-speed shaft of the gearbox creates a transistor-transistor logic (TTL) signal which is used as a once-per-revolution (OPR) trigger for the high frequency response data acquisition systems.

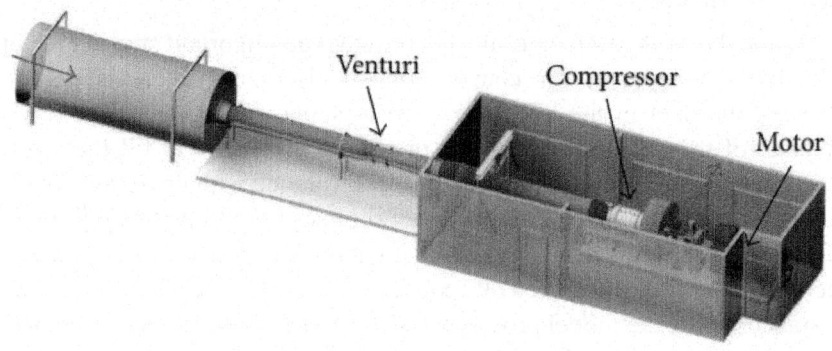

Figure 1: Purdue 3-stage axial compressor research facility.

The three-stage axial compressor features a 6061 aluminum casing and double-circular arc rotor blades made from 17-4 stainless steel. The rotor blade counts decrease through the compressor with 36 blades for Rotor 1, 33 blades for Rotor 2, and 30 blades for Rotor 3. The stator vanes in this facility are all shrouded without hub clearances. Additional information related to the airfoil geometry can be found in [14]. The compressor casing was machined with tight tolerances to create the best possible vehicle for tip leakage flow measurements. Specifically, the inner diameter of the casing was specified as 609.6 mm +0.025/−0.000 mm dimensionally, but additional geometric tolerances included a maximum 0.127 mm overall runout and a 0.051 mm surface profile shape. All dimensions were confirmed to be within tolerance using coordinate measuring machine (CMM) inspection techniques. Unless otherwise noted, the nominal rotor tip clearances for all data presented in this paper are 1.524 mm, representing 3.0% of the annulus height. For reference, Table 1 provides the aspect ratio of the rotor blades and the corresponding tip clearances as a function of average blade chord for the two tip clearance configurations referenced in this document.

Table 1: Rotor blade aspect ratios and clearance-to-chord values

	Rotor 1	Rotor 2	Rotor 3
Blade aspect ratio	0.76	0.72	0.68
Clearance-to-chord value (1.5% span)	1.14%	1.08%	1.02%
Clearance-to-chord value (3.0% span)	2.28%	2.16%	2.04%

Steady pressure and temperature measurements throughout the compressor are standard for all operations. These data include a seven-element radial distribution of stagnation pressure and stagnation temperature at the inlet and exit planes of the compressor (locations 0 and 9 in Figure 2), a circumferential and axial distribution of static pressures at all axial positions 0 through 9, and a series of surface-mounted T-type thermocouples on the outside of the compressor casing. These surface-mounted thermocouples are positioned at the aerodynamic interface plane (AIP) (plane 0 in Figure 2) and over each of the seven blade rows at a circumferential position of 124 degrees from the top of the compressor in the direction of rotor rotation, as shown in Figure 3. Past studies have shown that one surface temperature measurement is sufficient since circumferential surface temperature variations are less than 1°C. For data sets requiring interstage flow information, seven-element rakes measuring stagnation pressure and stagnation temperature are also inserted at each of the axial positions labeled 1 through 8 in Figure 2. A careful design of the temperature measurement system has reduced the calculated uncertainty to less than 0.3°C for all channels, and the uncertainty of the measured absolute pressures is less than or equal to 50 Pa.

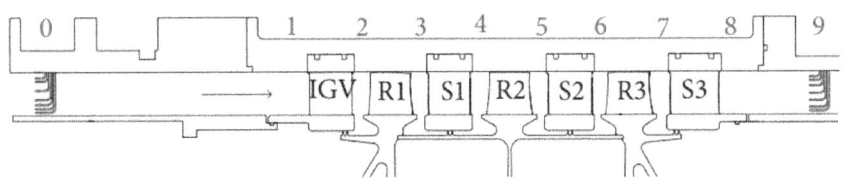

Figure 2: Compressor flow path and measurement plane locations.

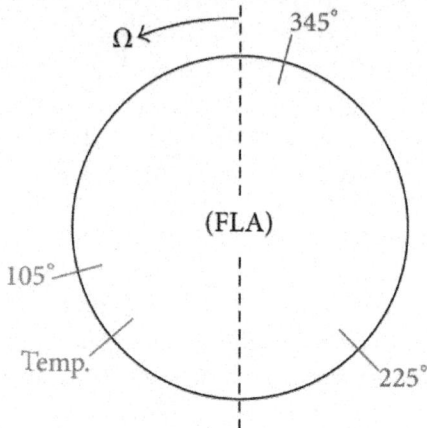

Figure 3: Circumferential instrumentation locations.

Capacitance Probe System

The capacitance probe system utilized for this study is a CapaciSense 5-series FM clearance measurement system produced by Pentair Thermal Management. There are nine channels available, allowing three probes to be implemented for each of the three compressor rotors. The three probes are equally spaced circumferentially, 120 degrees apart, at positions of 105 degrees, 225 degrees, and 345 degrees—all measured from the top of the compressor in the direction of rotor rotation. These locations are also marked in Figure 3. Each of the nine probes was individually calibrated using a custom-designed calibration disk which represents the tip geometry of the rotor blade through scaling techniques proven by the manufacturer. The probes were all calibrated for operation from a rub condition (0 mm clearance) to a maximum 5 mm clearance height.

The electronics chain for the FM tip clearance measurement system has several key components. The probes were designed and built for application in this specific facility; a triaxial cable is permanently attached to the probe and provides enhanced rejection of electromagnetic interference. The use of a non-mineral-insulated cable limits the maximum operating temperature of this system to 260°C. The oscillator drives the cable with an oscillating voltage (nominally 2 MHz). As the blade passes the probe, the measured capacitance modulates the driven frequency from the oscillator. This modulation is sensed by the carrier, and the demodulator converts the modulation frequency due to the blade passing event to a DC voltage. The DC voltage is correlated to a clearance height via the individual channel calibration. This proportionality between measured capacitance, frequency modulation, voltage, and tip clearance is the crux of the FM tip clearance measurement system.

The clearance measurement system is controlled from a set of "control and processing module" (CPM) computers. Each of these computers utilizes one Advantech PCI-1714UL data acquisition card capable of sampling at a rate of up to 10 MHz per channel, as well as an external trigger which is linked to the OPR signal from the TTL tachometer signal. For the measurements presented in this study, all data were collected using the full 10 MHz sampling capability of the data acquisition card. Other specifications for the capacitance probe measurement system are given in Table 2.

Table 2: Capacitance probe measurement system specifications

Parameter	Value
Operating frequency	2 MHz nominal
Oscillator sensitivity	100 kHz per pF
Demodulator sensitivity	500 mV per kHz typical
Measurement range	5 mm calibrated
System resolution	<0.001 mm (at 0.7 mm clearance)
Signal-to-noise ratio	30–50 typical

The clearance calculation process provides a blade-by-blade clearance output calculated from the peak-to-peak voltage for the typical blade pulse output signal (known as the blade passing signal (BPS)). As an alternative, the software applies a low-pass filter to the BPS output to create a DC voltage output signal which the manufacturer refers to as a "RMS" signal, although it should be noted that the low-pass filter mechanism does not represent a true root-mean-square calculation procedure. This RMS signal serves as a representative average measurement of the tip clearance for all blades by a particular probe. A separate calibration was performed for both the BPS and RMS output signals. All data presented in this study were calculated using data output from the RMS calculation procedure, but a comparison of the RMS with an arithmetic mean of BPS results (not shown here) agrees well.

UNCERTAINTY ANALYSIS

The manufacturer of the capacitance probe measurement system claims an umbrella uncertainty on the measurement system of less than 0.01 mm. This value is based on historical comparisons with other measurement techniques, including laser measurements and rub sticks, as well as careful attention to the design, manufacture, and calibration processes to ensure minimal uncertainties. However, no formal uncertainty analysis had yet been performed to validate this claim, which represents more of a repeatability or comparability and does not consider uncertainty contributions from the electronics components.

Other authors have performed uncertainty analyses for capacitance probe measurement systems, including Satish et al. [15] and Müller et al. [16]. The calculations from Satish et al. are extremely thorough as a reference for overall uncertainty analysis, but the manufacturer of the system used for this study did not provide sufficient information to perform the calculations outlined by Satish et al. The uncertainty analysis included by Müller et al. was performed for a comparable FM capacitance probe system several years ago and therefore was used as a model for the calculations performed here. Table 3 outlines some of the representative contributing components of the uncertainty analysis for the electronics of the capacitance probe system used for this study. Using these components, a root-sum-squared (RSS) calculation technique was implemented to calculate the overall system uncertainty. Table 3 includes the temperature coefficients for the demodulator and oscillator, but these components were not considered in the uncertainty analysis for two reasons: (i) the demodulator is operated in a climate-controlled environment to reduce thermal effects and (ii) the data presented in this paper were collected within 7°C of the calibration temperature, and convective cooling around the oscillators helps to reduce thermal effects related to temperature increases.

Table 3: Representative primary uncertainty analysis components for capacitance probe measurement system

Parameter	Uncertainty value
System noise	5 mV
Oscillator temperature coefficient	0.2% per °C
Demodulator temperature coefficient	0.1% per °C
A/D card uncertainty	1 mV
Standard deviation of calibration	80 mV
Probe setback uncertainty	0.0125 mm

The uncertainty of the electronics system is ultimately a function of the nominal clearance measured by the probe since the standard deviation of the calibration data varied inversely with increasing clearances. By calculating the representative uncertainty at each calibrated nominal clearance, a curve representing calculated system uncertainty versus nominal clearance was created for each of the nine probes, as shown in Figure 4(a). Because the probe-to-probe variability in Figure 4(a) is small (less than 0.005 mm) for nominal clearances of interest in this study (less than 2 mm), a representative average of the nine probes was calculated and a 4th-order polynomial fit was applied to those data, as shown in Figure 4(b).

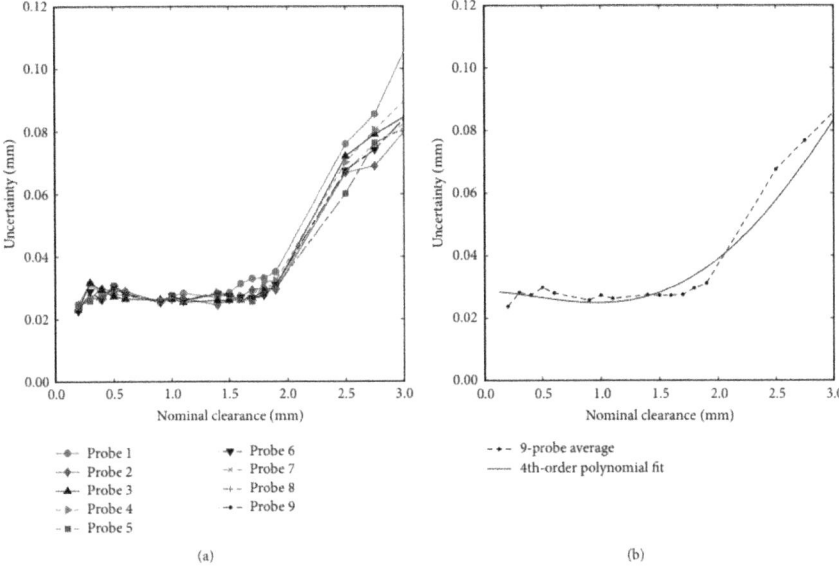

(a)　　　　　　　　　　　　　　　　(b)

Figure 4: Uncertainty of tip clearance measurements. (a) All nine probes; (b) average.

To further quantify systematic errors, a typical compressor build may include measurement of the static tip clearance over each rotor at each of the nine tip clearance probe locations using a micrometer and a dial indicator, as described by Brossman [17]; these measurements are then compared with the clearances calculated from the capacitance probe system at a low-speed operation (200 rpm is a typical limit for low-speed measurement) at the start of a compressor measurement campaign. By measuring the operating clearances immediately after start-up and at a low operating speed, influences due to centrifugal effects and thermal gradients can be considered negligible.

RESULTS

Rotational Speed

Figure 5 shows measured tip clearances as a function of the compressor operating speed during a typical startup process. These results are presented as a clearance difference with respect to the static clearance value:

$$\Delta\tau = \tau - \tau_{static}. \tag{1}$$

In Figure 5, the compressor is accelerated through a series of discrete mechanical speeds to allow the oil temperature in the gearbox manifold to increase to an appropriate level. During this process, the effect of tip clearance variation due to the increasing speed is apparent. However, tip clearance changes due to thermal effects are minimal since the temperature rise through the compressor is low at part speed, and the overall acceleration time is relatively small. As a result, the measured clearance changes in Figure 5 are largely due to the centrifugal effects imposed on the rotor as the rotor blades grow with increased operating speed. A minor exception to this is somewhat apparent for compressor operating speeds above 4500 rpm, when the measured clearances increase slightly during constant mechanical speed operation. At these higher-speed "plateaus," small contributions due to temperature rise through the compressor are first noticeable. However, because the compressor operates at an open throttle condition (near-choke) during the startup process, the temperature rise remains very low. These changes in tip clearance with increasing rotational speed have been documented by other authors in the past [8, 9, 16].

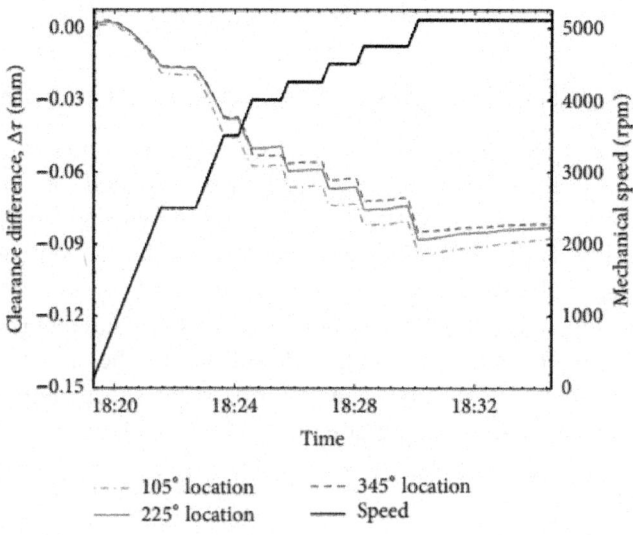

Figure 5: Clearance changes with speed for Rotor 2 during startup process at an open throttle operating condition.

Loading Condition

Once a steady operating speed has been reached, the loading of the compressor can be increased in the direction toward stall by closing the sliding-annulus throttle at the exit of the machine. The overall compressor total pressure ratio is shown in Figure 6 at each of the four corrected operating speeds. In Figure6, the total pressure is presented as a function of normalized inlet corrected mass flow rate, for which a normalized inlet corrected mass flow rate of unity denotes a nominal loading condition near the peak efficiency point. Each of the data points in Figure 6 represents a circumferential average of 20 vane positions with respect to measurement probes in order to remove variability due to stator wake effects. Each of the points in Figure 6 represents approximately one hour of compressor operating time. Understandably, the temperature rise through the compressor increases with increased loading. As this occurs, the thermal effects begin to have a greater impact on the outer casing of the compressor. Specifically, moving from the open throttle position to a near-stall operating condition has the effect of increasing the stagnation temperature rise at the exit of each rotor blade row due to the increased work done on the flow.

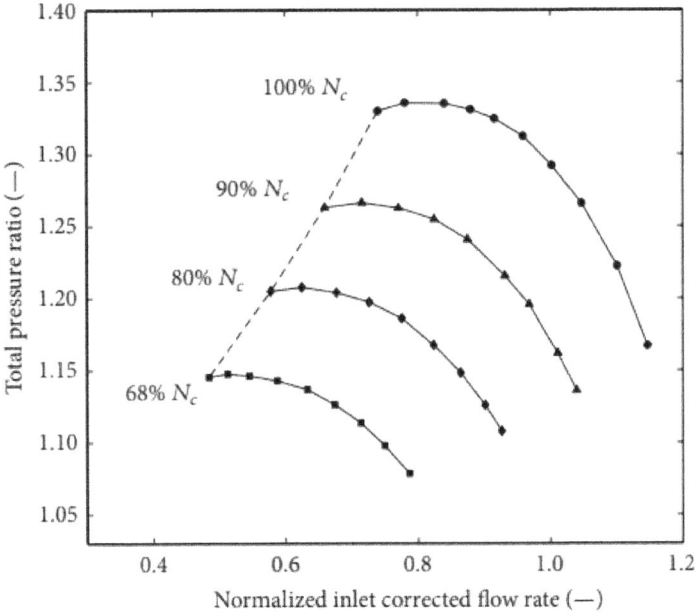

Figure 6: Overall compressor map at four operating speeds.

Figure 7 shows the measured tip clearances for each of the nine loading conditions on the 100% corrected speedline from the compressor map in Figure 6. During the measurement of the overall compressor performance map in Figure 6, the operating clearances were measured, but since each data point in Figure 6corresponds to approximately one hour of compressor operation, there is a possibility that any observed trends in measured clearance with loading condition are affected by variations in ambient temperature. In an effort to avoid false conclusions due to this potential thermal affect, the rotor clearances shown in Figure 7were measured at the same loading conditions corresponding to the 100% corrected speedline from Figure 6. However, the data in Figure 7 were collected with approximately ten minutes between adjacent points during a period of time when the ambient temperature was relatively constant (within 2°C). This time allows the machine to sufficiently reach a thermal equilibrium without allowing ambient temperatures to vary significantly. Furthermore, Figure 7 presents two series of measured clearances collected on separate occasions to represent the repeatability of the measurements as an alternative to the calculated electronic system uncertainty from Figure 4. With few exceptions, Figure 7 shows measurement repeatability of 0.01 mm or better, which is in agreement with the value reported by the system manufacturer.

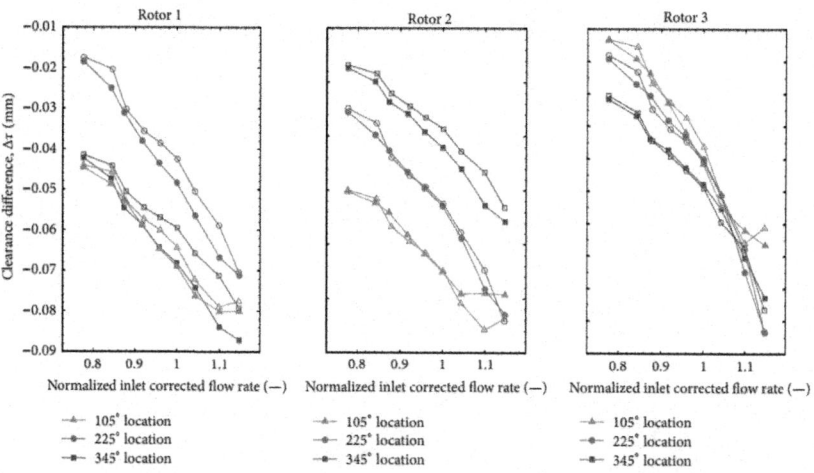

Figure 7: Clearance changes with compressor loading condition at a 100% corrected operating speed.

The change in tip clearance with compressor loading condition has been predicted by Dong et al. [12] using a series of model equations for clearance change due to thermal and centrifugal effects. Predicted clearances calculated using this model were verified using experimental measurements collected from the General Electric E^3 ten-stage compressor build. For a separate 11-stage compressor, Dong et al. predicted that increasing the loading condition from near choke to near-stall would affect the tip clearance by approximately 0.1% span for Rotor 4 but as much as 0.4% for Rotor 10 (due to an increase in temperature rise in the rear stages of the compressor).

The model presented by Dong et al. [12] utilizes a series of six calculated deformation contributions, each of which can be calculated using independent equations and combined to calculate the overall clearance. Specifically, the model calculates blade deformation due to thermal expansion and centrifugal forces, shroud deformation due to thermal expansion and pressure forces, and disc deformation due to thermal expansion and centrifugal forces. The six contributions are then combined with the static assembled clearance to determine the model clearance

$$\tau_{model} = \tau_{static} - \left(u_{B,thermal} + u_{B,centrif}\right)$$
$$+ \left(u_{S,thermal} + u_{S,pressure}\right) - \left(u_{D,thermal} + u_{D,centrif}\right),$$

(2)

where the subscripts B, S, and D represent the blade, shroud, and disc, respectively. The equations representing these six contributing deformations have been applied using pressure and temperature data corresponding to the compressor performance points presented in Figure 6 for comparison with operating rotor tip clearances measured simultaneously with the capacitance probe system.

The equations for the blade and shroud deformations were applied exactly as suggested by Dong et al. [12]. However, the equation for the disc deformation due to thermal effects was simplified to accommodate information available from the experimental results. Specifically, there were insufficient experimental data to allow calculation of the radial temperature distribution on the disc. As a result, the equations were simplified to approximate the disc temperature distribution by a constant temperature measured at the inner diameter of the flow path. This approximation is based on two primary pieces of information: (i) radial temperature gradients in the disc are expected to be small for the front stages of axial compressors and (ii) the integrated-bladed-rotor (blisk) design of the Purdue three-stage axial compressor does not have air paths to facilitate convective cooling.

Using the series of stagnation pressures, stagnation temperatures, and static pressures in the flow path combined with temperatures measured on the outside of the casing, the six components required for the Dong et al. model were calculated. In the cases when static temperatures were required, the thermodynamic equation program REFPROP [18] was utilized in combination with measured parameters, allowing the application of high-accuracy equations to calculate thermodynamic properties including any pertinent humidity effects. Specifically, measured stagnation temperatures and pressures were combined with measured static pressure at the wall (assuming a constant radial distribution of static pressure due to a constant annulus height) to calculate the required static temperatures.

The results comparing the experimental data with the calculated model results are shown in Figures 8 and 9for 100% and 90% corrected operating speeds. In this case, the results are presented as a difference with respect to the open throttle loading condition, denoted by the subscript OT:

$$\Delta \tau_{OT} = \tau - \tau_{OT}. \tag{3}$$

Considering first the results of the 100% corrected speedline data shown in Figure 8, there is a noticeable lack of agreement between the measured clearances and the predicted clearances for normalized inlet corrected mass flow rates less than one. Specifically, for the results from Rotor 2 in Figure 8, the near-stall point (lowest mass flow rate) shows a discrepancy nearly two times the uncertainty presented in Figure 4 and at least five times the repeatability shown in Figure 7. On the other hand, the 90% corrected speedline data in Figure 9show excellent agreement between the predicted and measured clearances. Similar agreement between predicted and measured clearances also exists for the 80% and 68% corrected speedlines (not shown here).

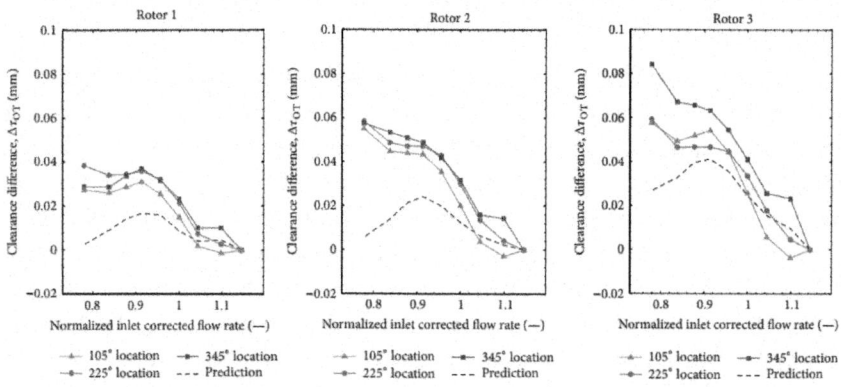

Figure 8: Predicted and measured clearance variations with loading condition, presented as a difference from the open throttle condition (100% N_c).

The model equation presented in (1) suggests that the discrepancy for the 100% corrected speedline results could come from an overprediction of blade and/or disk growth or an underprediction of shroud growth. To more closely consider the source of the errors, the individual components are presented in Figure 10. From this figure, the contributions to the predicted clearances are driven primarily by the thermal growth, which yield clearance differences several orders of magnitude larger than the pressure or centrifugal components.

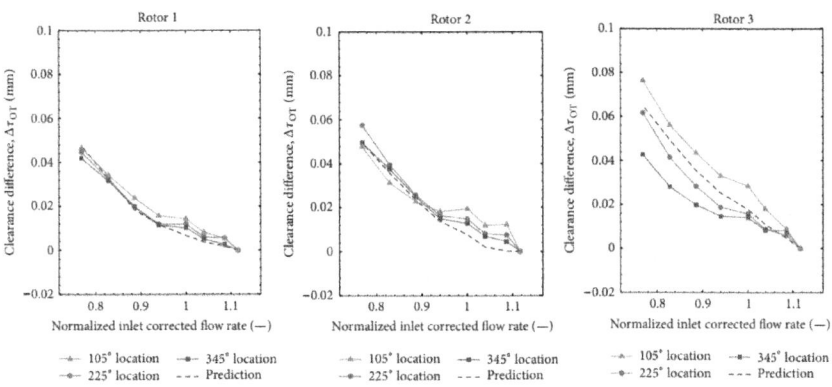

Figure 9: Predicted and measured clearance variations with loading condition, presented as a difference from the open throttle condition (90% N_c).

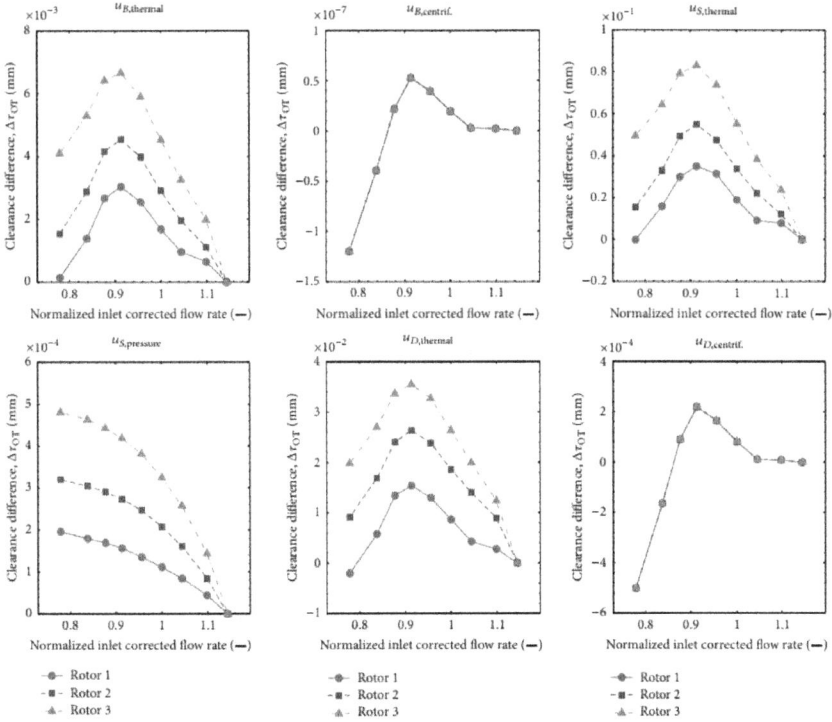

Figure 10: Components of predicted clearances from (2) corresponding to data for Figure 8.

Ambient Temperature Effects

Analysis of the data corresponding to the results in Figure 8 shows that a significant temporal variation of ambient temperature (on the order of 3°C) exists for the data at low inlet corrected mass flow rates. These variations of temperature are present throughout the data collection period due to changes of ambient conditions. Of course, the corrected parameters defining the operating point (rotational speed, mass flow rate) appropriately account for these varying ambient conditions in overall performance, but the averaged data which were utilized to calculate the temperatures required for use in (2) do not appropriately represent the required parameters. Thus, an adjustment was made to the prediction calculations such that the temperature utilized for the thermal components of (2) represents the temperatures measured at the exact same time as the measured tip clearances. Figure 11 shows that the adjustment does not remedy the entire discrepancy, but it more closely represents the measured values on the order of the uncertainty of the measurements. No adjustments were made to the nonthermal components of (2) due to their relatively small contribution shown in Figure 10.

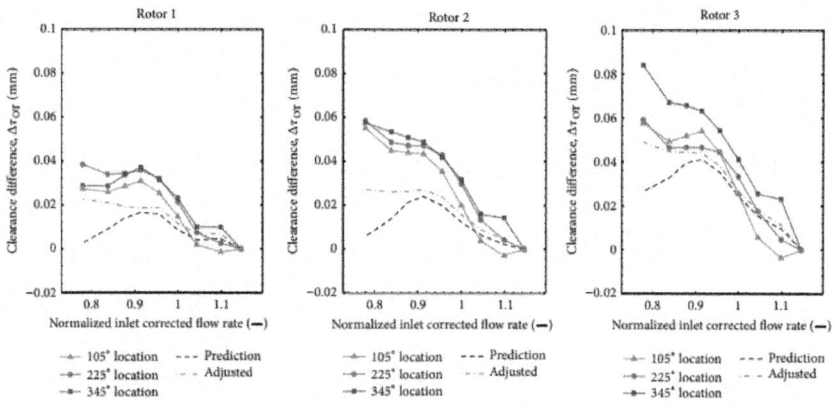

Figure 11: Predicted and measured clearance differences from the open throttle loading condition including a temperature adjustment for the predicted results (100%N_c).

Careful consideration is always given to ensure that a thermal equilibrium is reached prior to collecting these data. However, the data presented thus far suggest that the clearance variations with loading condition can also depend on the ambient conditions. Figure 12 shows this by comparison of two different tip clearance cases corresponding to different ambient conditions. Specifically, the data presented previously for the 3.0% clearance was susceptible to variations of ambient temperature for the low flow rates (as discussed above), whereas the 1.5% tip clearance results were collected over a period of time when the ambient temperature was nearly constant, which leads to the increased clearance growth at the low flow rates. The change of nominal tip clearance between the two cases poses a potential source of variability, particularly in terms of temperatures, but these data were collected during periods of similar weather and previous studies have shown that the temperature rise through the compressor is nearly identical for these two tip clearance cases.

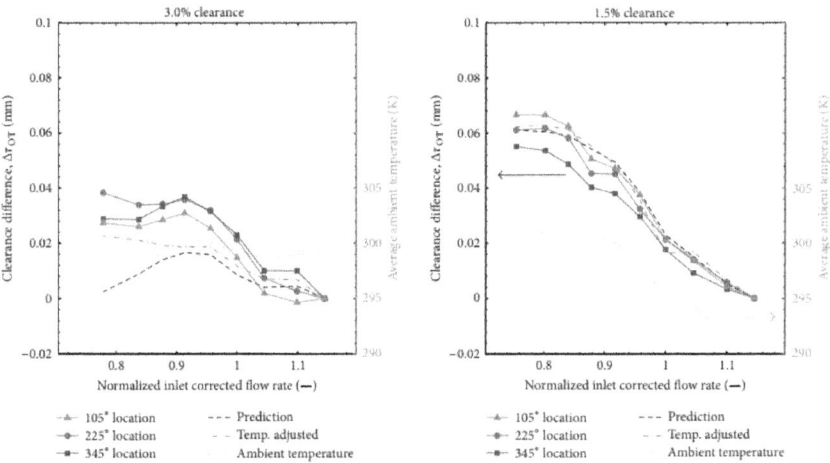

Figure 12: Predicted and measured clearance variations with loading condition, presented as a difference from the open throttle condition for different compressor tip clearance configurations with variations due to ambient conditions (100% N_c, Rotor 1).

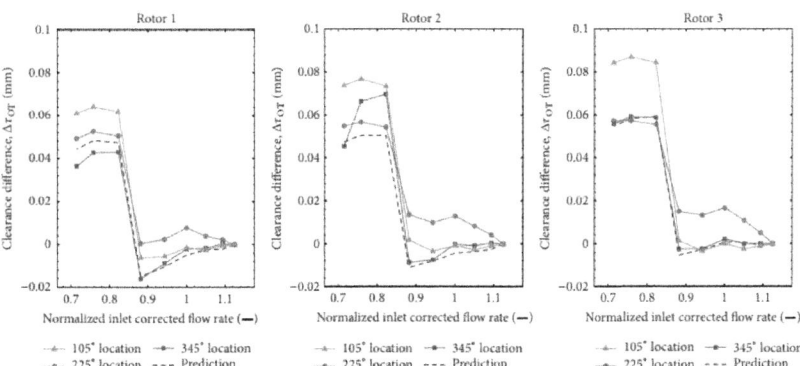

Figure 13: Predicted and measured clearance variations with loading condition, presented as a difference from the open throttle condition (80% N_c) for compressor testing with a "stop and start" (1.5% tip clearance).

To further emphasize the effect of ambient temperature on measured rotor tip clearance, Figure 13 shows the results of the 80% corrected speedline for one of the tip clearance configurations. It is important to note here that the lower rotational speed creates less temperature rise through the compressor. This lower temperature rise leads to clearance measurements which are less sensitive to loading condition. In other words, the measured and predicted clearances for adjacent operating points are nearly constant.

Considering the data in Figure 13, a stretch of severe weather forced a stop of the experimental campaign after the completion of the three points with the lowest flow rates, and the data collection process resumed on the following day. As shown in Figure 13, the discontinuity of ambient conditions for between the third and fourth points in the order of increasing flow rates caused a discernable discontinuity of rotor tip clearance (measured and predicted). This jump in clearance is approximately 0.06 mm, which represents an 8% change of operating tip clearance for the nominal tip clearance of 1.5% span (0.762 mm), and is attributable to a change in ambient temperature of approximately 15°C. Although there is no measureable change of performance parameters due to this difference of rotor tip clearance in this compressor, these results suggest that it is advisable to avoid a "stop and start" when a comparison of tip clearance information between adjacent data points is desired.

In these cases of varying ambient temperature and other environmental considerations, it is possible that the changing air composition could affect the dielectric properties enough to skew the capacitive effect, thereby artificially introducing the measured clearance changes shown here. Chivers [9] analyzed the relative permittivity of the dielectric material (for this study, air) at several temperatures and pressures representing a 220 kN thrust class gas turbine engine. From that analysis, Chivers showed that increases in pressure and temperature due to flow through the compressor would affect the relative permittivity of air by approximately 0.05% compared to the same properties at the inlet of the machine. Further, Baxter [19] addresses the concern of humidity effects on relative permittivity. An increase of relative humidity from 40% to 90% at 20°C (a representative average for the data presented here) would affect the relative permittivity and, therefore, the measured capacitance on the order of 0.007%—an order of magnitude less than changes due to temperature alone. From these values, it is expected that the changes of dielectric properties do not have a significant effect on the measured clearances.

Thermal Equilibrium Considerations

To this point, analysis of rotor tip clearance effects due to rotational speed, loading condition, and ambient temperature has been considered. Allusion has been given to thermal equilibrium, but primarily in the framework of ambient temperature variations due to time of day. Thus, the time associated with approaching a condition of equilibrium based on measured clearances was also considered explicitly in the context of thermal equilibrium. The compressor was operated at a steady rotational speed of 2500 rpm and was allowed to reach a state of equilibrium as determined from measured thermocouples and tip clearance measurements. At that point, the operating speed was increased to 5150 rpm (a mechanical speed relating to a 100% corrected rotational speed based on ambient conditions) at a linear rate of 1100 rpm per minute, and the corresponding response of measured clearances and shroud outer diameter temperatures was observed. In addition to the mechanical speed in Figure 14(d), the measured ambient temperature is also shown to verify that the measured temperature and clearance changes in Figures 14(a)–14(c) are not a result of changes in ambient temperature. These clearance results are shown in Figures 14(a)–14(c) as a difference with respect to the static clearance values.

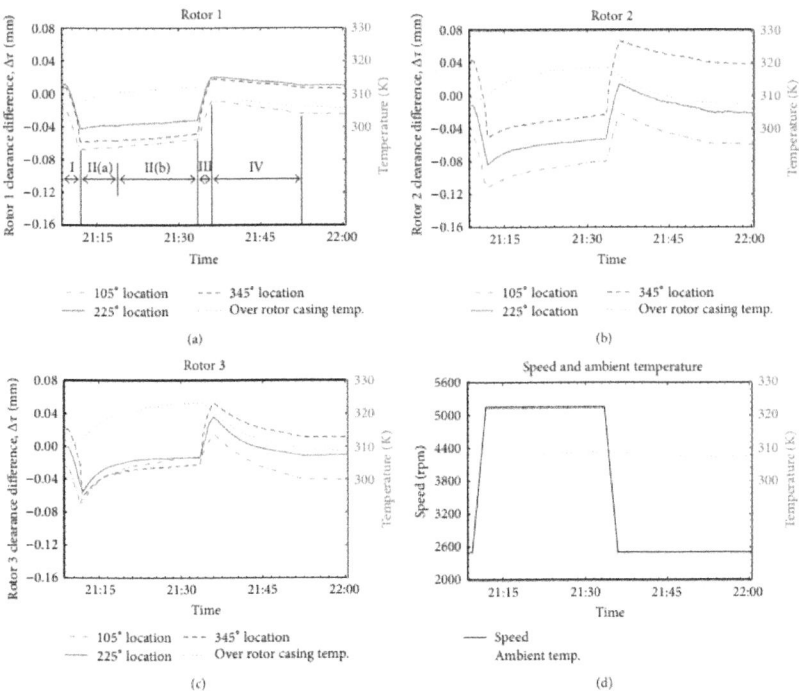

Figure 14: Clearance changes due to compressor speed changes and corresponding thermal variations at a nominal loading condition.

Referring to Figure 14, immediately after the increase of speed, the measured clearances decrease in a nonlinear fashion. Throughout this speed change process, the surface temperatures measured at the outer diameter of the compressor shroud show no change. In this period (Region I; approximately 2-3 minutes), an increase of flow temperatures exists, but there is a definitive lag in the clearance response due to thermal growth. Over the course of the following 15 minutes (Region II), however, the temperature increase measured at the outer diameter of the compressor shroud is observable for all three rotors following an asymptotic trend. During this same period, the measured clearances also show an increase, but the trends for the three rotors differ slightly.

Throughout the entirety of Region II, the clearance change over Rotor 1 is approximately linear. In contrast, the clearance changes over Rotor 2 and Rotor 3 exhibit a nonlinear section, Region II(a), which extends across approximately the initial one-third of Region II. The nonlinear response of Region II(a) is expectedly due to the more immediate change of tip clearance due to temperature rise in the flow (e.g., the contribution $u_{B,thermal}$). For this reason, the clearance change in Region II(a) for Rotor 3 is more significant than that of Rotor 2, and the relatively small temperature rise across Rotor 1 leads to a nondiscernable Region II(a) for that blade row. Region II(b), on the other hand, is primarily a contribution due to the heat transfer through the compressor shroud contributing to the thermal growth term $u_{S,thermal}$.

Once the measured temperatures at the outer diameter of the shroud approached an asymptotic value, the rotational speed was decreased at the same rate of 1100 rpm per minute. Predictably, the trends of measured clearances show a sharp increase with the decreased rotational speed as the clearance contributions due to centrifugal forces respond (Region III). In contrast with Region II, the thermally dominated Region IV for the deceleration follows a more linear trend for all three rotors. This important difference is related to the cooling process imposed by decreasing speed as the temperature rise through the compressor decreases and the convective heat transfer to the air around the compressor increases. As a result, there is a less distinct separation of thermal contributions in Region IV. Although there is a slight asymptotic trend in the clearance variations of Region IV for Rotor 3, the trend is significantly less apparent than it was for Region II with the same blade row.

SUMMARY AND CONCLUSIONS

Operating rotor tip clearances have been measured using an FM capacitance probe measurement system with the Purdue three-stage axial compressor. Due to the large amount of experimental data available, this facility provides a unique opportunity to examine tip clearance variations due to several external factors. In particular, tip clearance changes due to rotational speed, loading condition, ambient temperature, and thermal equilibrium have been addressed. This study has experimentally validated the existence of measurable changes in tip clearance at several loading conditions and for several different compressor tip clearance heights. Many of these measured changes are well beyond the calculated uncertainty of the electronics system, and nearly all are beyond the repeatability of the measurements.

The results presented herein pose a very real problem for operating gas turbine engines that has not been presented previously through experimental data. Previous authors have discussed the potential variation of rotor tip clearance with changing loading condition, but this changing clearance is typically not implemented in computational models. These measurable changes of tip clearance with operating condition could be important for comparing experimental and computational performance. More importantly, these measured clearance changes with ambient operating conditions can also lead to measurable changes in overall total pressure ratio beyond the uncertainty of the measurements.

These results have focused on a machine which yields relatively low total temperature ratios. Thus, the thermally driven clearance changes could be considered small in comparison with larger multistage compressors. In particular, it has been shown that the largest clearance variations exist in the rear stages where the temperature rise is the greatest. For typical gas turbine compressor designs, this is the location at which the blade heights and the relative tip clearances are the smallest. As a result, the change of tip clearance with operating condition and ambient temperature is likely more significant than some may believe.

Ultimately, the results presented here motivate a need to incorporate tip clearance measurement systems with compressor test facilities desiring to accurately monitor compressor performance. Experimental data collected at different times of day, during different periods of the calendar year, or at different loading conditions can lead to significant variations of tip clearances which must be monitored for a best comparison with computational tools.

ACKNOWLEDGMENTS

This material is based upon work supported by NASA under the ROA-2010 NRA of the Subsonic Fixed Wing project and in part by the National Science Foundation Graduate Research Fellowship Program under Grant no. DGE-1333468 The authors are grateful to Reg Morton for providing technical information specific to the probe measurement system. The authors would also like to thank Rolls-Royce for the permission to publish this work.

REFERENCES

1.   D. C. Wisler, "Loss reduction in axial-flow compressors through low-speed model testing," ASME Journal of Engineering for Gas Turbines and Power, vol. 107, no. 2, pp. 354–363, 1985. · ·

2.   C. C. Koch and L. H. Smith Jr., "Loss sources and magnitudes in axial-flow compressors," ASME Journal of Engineering for Gas Turbines and Power, vol. 98, no. 3, pp. 411–424, 1976. · ·

3.   J. P. Barranger, "Recent advances in capacitance type of blade tip clearance measurements," NASA Technical Memorandum 101291, 1998.

4.   J. D. Denton, "Some limitations of turbomachinery CFD," ASME Paper GT2010-22540, American Society of Mechanical Engineers, 2010.

5.   P. P. Walsh and P. Fletcher, Gas Turbine Performance, Blackwell Science, Oxford, UK, 2008.

6.   N. R. Smith, R. A. Berdanier, J. C. Fabian, and N. L. Key, "Reconciling compressor performance differences with varying ambient conditions," ASME Paper POWER2015-49102, American Society of Mechanical Engineers, 2015.

7.   D. E. van Zante, A. J. Strazisar, J. R. Wood, M. D. Hathaway, and T. H. Okiishi, "Recommendations for achieving accurate numerical simulation of tip clearance flows in transonic compressor rotors," ASME Journal of Turbomachinery, vol. 122, no. 4, pp. 733–742, 2000. · ·

8.   G. Sheard, "Blade by blade tip clearance measurement," International Journal of Rotating Machinery, vol. 2011, Article ID 516128, 13 pages, 2011. · ·

9.   J. Chivers, "A technique for the measurement of blade tip clearance in a gas turbine," AIAA Paper 89-2916, AIAA, 1989.

10.   H. Agarwal, S. Akkaram, S. Shetye, and A. McCallum, "Reduced order clearance models for gas turbine applications," AIAA Paper 2008-2177, AIAA, 2008.

11.   J. A. Kypuros and K. J. Melcher, "A reduced model for prediction of thermal and rotational effects on turbine tip clearance," NASA Technical Memorandum 2003-212226, 2003.

12.   Y. Dong, Z. Xinqian, and L. Qiushi, "An 11-stage axial compressor performance simulation considering the change of tip clearance in different operating conditions," Proceedings of the Institution of Mechanical Engineers Part A: Journal of Power and Energy, vol. 228, no. 6, pp. 614–625, 2014. ·

13. D. C. Ye, F. J. Duan, H. T. Guo, Y. Li, and K. Wang, "Turbine blade tip clearance measurement using a skewed dual-beam fiber optic sensor," Optical Engineering, vol. 51, no. 8, Article ID 081514, 2012. · ·

14. N. L. Key, "Influence of upstream and downstream compressor stators on rotor exit flow field,"International Journal of Rotating Machinery, vol. 2014, Article ID 392352, 10 pages, 2014. ·

15. T. N. Satish, R. Murthy, and A. K. Singh, "Analysis of uncertainties in measurement of rotor blade tip clearance in gas turbine engine under dynamic condition," Proceedings of the Institution of Mechanical Engineers, Part G: Journal of Aerospace Engineering, vol. 228, no. 5, pp. 652–670, 2014. · ·

16. D. Müller, A. G. Sheard, S. Mozumdar, and E. Johann, "Capacitive measurement of compressor and turbine blade tip to casing running clearance," ASME Journal of Engineering for Gas Turbines and Power, vol. 119, no. 4, pp. 877–884, 1997. · ·

17. J. R. Brossman, An investigation of rotor tip leakage flows in the rear-block of a multistage compressor [Ph.D. thesis], Purdue University, West Lafayette, Ind, USA, 2012.

18. E. W. Lemmon, M. L. Huber, and M. O. McLinden, NIST Standard Reference Database 23: Reference Fluid Thermodynamic and Transport Properties-REFPROP, Version 9.1, National Institute of Standards and Technology, Standard Reference Data Program, Gaithersburg, Md, USA, 2013.

19. L. K. Baxter, Capacitive Sensors: Design and Applications, Wiley-IEEE Press, Hoboken, NJ, USA, 1996.

Chapter 12

A MINIATURE FOUR-HOLE PROBE FOR MEASUREMENT OF THREE-DIMENSIONAL FLOW WITH LARGE GRADIENTS

Ravirai Jangir, Nekkanti Sitaram, and Ct Gajanan

Thermal Turbomachines Laboratory, Department of Mechanical Engineering, IIT Madras, Chennai 600 036, India

ABSTRACT

A miniature four-hole probe with a sensing area of $1.284\,mm^2$ to minimise the measurement errors due to the large pressure and velocity gradients that occur in highly three-dimensional turbomachinery flows is designed, fabricated, calibrated, and validated. The probe has good spatial resolution in two directions, thus minimising spatial and flow gradient errors. The probe is calibrated in an open jet calibration tunnel at a velocity of 50 m/s in yaw and pitch angles range of ±40 degrees with an interval of 5 degrees. The calibration coefficients are defined, determined, and presented. Sensitivity coefficients are also calculated and presented. A lookup table method is used to determine the four unknown quantities, namely, total and static pressures and flow angles. The maximum absolute errors in yaw and pitch angles are 2.4 and 1.3 deg., respectively. The maximum absolute errors in total, static, and dynamic pressures are 3.4, 3.9, and 4.9% of the dynamic pressures, respectively. Measurements made with this probe, a conventional five-hole probe and a miniature Pitot probe across a calibration section, demonstrated that the errors due to gradient and surface proximity for this probe are considerably reduced compared to the five-hole probe.

INTRODUCTION

The use of the multihole pressure probes has become common to determine total and static pressures, flow velocity, and flow directions in three-dimensional flow fields with suitable calibrations. Multihole pressure probes make accurate and simultaneous measurement of total and static pressures and flow direction when pressure and velocity gradients are small. Pressure probes have some advantages over other methods as their maintenance, relatively low cost, and simplicity in operation. Hence these are preferred in research and industrial purposes. In principle, any aerodynamic body such as cylinder, sphere, wedge, or prism, with a number of holes can be used to measure three-dimensional flows. A minimum of four holes on an aerodynamic body is required to measure the four unknowns, namely, total and static pressures and two angles in mutually perpendicular planes, in three-dimensional flows. However for the sake of symmetry and extended range of measurement capability, five-hole and seven-hole probes are preferred.

Because of their simplicity in operation and low cost, extensive investigations are carried out on multihole probes, particularly on five-hole probes, on their calibration and data reduction methods and their application to complex three-dimensional flow measurements. Treaster and Yocum [1] reported on calibration of different types of five-hole probes and their errors due to Reynolds number variation and surface proximity effects. Pisasale and Ahmed [2] had developed a method to extend the useful operating range for highly three-dimensional flows by replacing the conventionally defined denominator in the calibration coefficients with a more complex denominator. Pissasale and Ahmed [3, 4] developed theoretical relationship based on potential flow for calibration and application of five-hole flows. Yasa and Paniagua [5] developed a robust method for five-hole probe calibration. This technique demonstrates that the five-hole probe can be used even if one of the side holes is blocked. Dominy and Hodson [6] investigated the effect of various factors including head shape and Reynolds number on the calibration of five-hole probes. Lee and Jun [7] calibrated a commercial five-hole probe at different Reynolds numbers. They found that the effect of Reynolds number on the calibration coefficients is different at different yaw and pitch angles. A comprehensive review of recent developments in multihole probe technology is presented by Telionis et al. [8]. Recently Lien and Ahmed [9] have used a five-hole probe to measure skin friction coefficient in complex two and three-dimensional flows. This technique avoids the necessity of aligning a Preston probe with the flow direction and the necessity of a wall static tap.

But in three-dimensional flows with large pressure and velocity gradients, in flows such as tip clearance vortex and other complicated flow phenomena that occur in turbomachinery, these probes make erroneous measurements due to their relatively large size.

Spatial errors can be minimized in two ways first by minimizing the probe head dimensions and second by applying corrections. Earlier, work was done on the miniaturization of multihole probes. To characterize Dean's vortices, Ligrani et al. [10] developed a miniature five-hole probe (diameter 1.22 mm) and used in low speed channel flow and they applied a correction method to account for spatial errors. However this method has limitations on the size of the pressure tubes that are used to make the probes. Smaller diameter tubes have longer response time. Also the small tubes may be easily blocked by dirt and may give erroneous measurements or may not give any measurements.

The other alternative is to develop methods to correct the measurements for the errors due to pressure and velocity gradients and surface proximity. This approach was adopted by Chernoray and Hjärne [11], Town et al. [12], and Honen et al. [13]. However these methods may require larger number of measurements. For the minimum errors due to pressure and velocity gradients and surface proximity effects, it will be necessary to combine both the techniques.

The probe size is relatively large in the above cases. However the pressure and velocity gradients are large in tip clearance vortex, end wall flows, flow in corner of blades, and other complex flows that occur in turbomachinery. Hence there is a strong requirement of further miniaturisation of the probe head for highly three-dimensional flow measurements. For three-dimensional flow measurements a four-hole probe which can measure four independent pressures can be used.

Four-hole probes have some advantages over five- and seven-hole probes as fewer measurements and reduced instrumentation are required during calibration and application. The measuring volume of the probe head is small compared with the five-hole probe. Hence the spatial errors caused by large pressure and velocity gradients and errors due to the surface proximity effects and shear gradients effects are reduced.

Four-hole probes come in many configurations and are used for many applications. The simple and earliest four-hole probe was obtained by modifying a three-hole cylindrical probe. An additional hole which is mainly sensitive to the flow in the pitch plane is added to the end of a three-hole probe. This type of probe is known as cantilevered four-hole cylindrical probe and used in many measurement applications in turbomachinery (Erwin, [14]) and other flows (Maheshwari et al., [15]). Similar four-hole probe with wedge configuration is commercially available from AC-flow Corporation [16]. Heneka [17] and

Ainsworth et al. [18] developed similar four-hole wedge probes with fast response pressure transducers to measure periodic total and static pressures, velocity and its three components, and flow angles. These probes are usually large about 3 mm in diameter and the measuring errors due to pressure and velocity gradients and surface proximity are usually large. However it has to be mentioned that Schlienger [19] developed a cantilever cylindrical probe of 1.2 mm tip diameter with a spatial resolution of about 1 mm. This type of probe is very useful to measure the flows in diffusers of centrifugal compressors and in labyrinth seals of axial turbines, where the spanwise flow angles are usually small. However for many turbomachinery flow measurements, pressure probes with very small measurement volumes and capability to measure large spanwise angles are needed. The available literature on such probes is presented below.

A four-hole probe which satisfies partially the above requirements was developed by Shepherd [20] for the three-dimensional flow measurements. The main feature of the probe is a tip shaped like the frustrum of a pyramid, with three-side holes equispaced around a central hole. This probe was calibrated in yaw and pitch angle range of ±40° and the calibration space is divided into six zones, making the use of this probe somewhat complicated. Sitaram and Treaster [21] have developed and presented two miniature four-hole probes. The probe heads were fabricated from 0.55 mm outer diameter and 0.30 mm inner diameter stainless steel hypodermic tubes. The probe heads are machined to a 50 degree half angle cone and located approximately four local support diameters upstream to reduce support interference effects. These probes have slightly less measuring volumes and higher spatial resolution compared to the probe developed by Shepherd. The probe with pyramid head was extensively used in many flow measurement applications.

Based on the above literature the commonly used four-hole probes can be divided into three designs, namely, modified cylindrical/hemi spherical probe, wedge/pyramid probe, and forward facing tube probe. These probes are presented in Figure 1. Both modified cylindrical/hemi spherical probe and wedge/pyramid probe can be used when the measurement space is limited, such as in diffuser passages of centrifugal compressors and turbomachinery seals. But the pitch angle range of these probes is limited and cannot be increased. A properly designed forward facing tube four-hole probe can be used to measure highly three-dimensional flows with large variations in both yaw and pitch angle ranges.

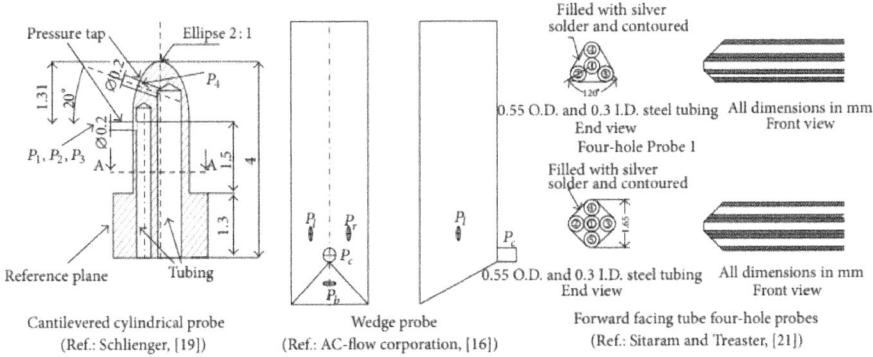

Figure 1: Commonly used four-hole probe configurations.

The forward facing tube probe 2 shown in Figure 1 can be used both in the four-hole and five-hole probe modes. In the four-hole probe mode, tube 5 was not used. Tube 5 could be eliminated or used to mount a thermocouple. This probe has a smaller error due to velocity gradient effect, as the probe height in the pitch plane is equal to two-tube diameters only. The four-hole probe 1 shown in 1 and five-hole probe have larger errors due to velocity gradient effect, as their heights in the pitch plane are two and one-half times tube diameters and three times tube diameters, respectively. It should be noted that these probes are smaller than any commercially available probes, so that spatial and velocity gradient effects are minimized.

The probes described above have small errors due to pressure and velocity gradients and surface proximity errors. However these errors are still large in many turbomachinery flows. Hence it is necessary to develop methods to further reduce the spatial and surface proximity errors.

The four independent pressures can be related to determine the four unknowns, namely, total and static pressures and two flow angles in mutually perpendicular planes. From the total and static pressures, velocity can be calculated. Using the flow angles, the three velocity components can be calculated. Hence a four-hole probe is chosen for development. Probe tip design can vary diversely depending on the particular application. A pressure probe must be designed for particular application where it is used. Four-hole probes are used by Sitaram and Treaster [21] and other researchers. However the configuration used by these researchers has good spatial resolution in one direction only. A four-hole probe with good spatial resolution in two directions for highly three-dimensional flow measurement is developed in this paper.

OBJECTIVE

The objective of the present investigation is to design, fabricate, calibrate, and validate a miniature four-hole probe with minimum spatial error for accurate three-dimensional flow measurement with large pressure and velocity gradients as in cases of tip clearance vortex, flows in corner of blades, end wall flows, and other complex flows that occur in turbomachinery.

DESIGN AND FABRICATION OF FOUR-HOLE PROBE

For three-dimensional flow, four (at least) or more holes strategically located on a probe head are necessary to determine the flow. Design of any multihole probe is a compromise between many conflicting requirements, such as small probe tip versus good response and large yaw and pitch angle measurement capability versus sensitivity of the calibration coefficients. As the size of the probe tip is reduced, by using smaller tubes, the response of the pressure measuring time increases. Hence more time is required for data acquisition. Smaller chamfer angle of the probe gives larger yaw and pitch angle measurement capability at the expense of reduced sensitivity of the calibration coefficients. The present probe is designed to optimise these conflicting requirements. In the present investigation, the four holes correspond to the centre hole, one yaw hole and two pitch holes of a fivehole probe. This design deviates from the design of earlier four-hole probes, which use one centre hole, two yaw holes, and one pitch hole of a five-hole probe. The commonly used five-hole probes are symmetric about both yaw and pitch axes, while the commonly used four-hole probes are symmetric about the yaw axis. The commonly used fourhole probes have more yaw angle range and better sensitivity of calibration coefficients in the yaw plane. The advantage of the present design is that the pitch angle range can be increased, although the yaw angle range is reduced.This is not a major disadvantage as the probe can be easily change its yaw angle when the flow exceeds the calibration yaw angle range. However a new technique needs to be developed to determine reference yaw angle. The centre hole with a chamfer angle of 90∘ to the tube gives a measure of total pressure and is denoted by P_C. The side hole is chamfered at an angle of about 35∘ to the yaw plane and is noted as P_S and is mainly sensitive to the yaw angle variation. The bottom and top holes are chamfered at an angle of about 35∘ to the pitch plane and are noted as P_B and P_T, respectively, and are mainly sensitive to the pitch angle variation. Hence four independent pressures can be measured, which are sufficient to define the threedimensional flow. It has been already demonstrated that small chamfer angle gives higher calibration angle range. Hence small chamfer angle is chosen for the present design

The probe head design is chosen so as it has good spatial resolution in both yaw and pitch directions. Assuming the tube size is the same, five-hole and seven-hole probes have a spatial resolution of 3 (: tube diameter) in both directions. Four-hole probes of existing designs have a spatial resolution $3d$ (d: tube diameter) in both directions. Four-hole probes of existing designs have a spatial resolution of $3d \times 2d$ in yaw and pitch directions. The present probe has a spatial resolution of $2d$ in both directions. Therefore it can be used for highly three-dimensional flows such as flow in corner of blades and end wall flows with large pressure and velocity gradients in all directions. The probe is L-shaped and the probe head details are shown in Figures 2(a) and 2(b). The material of the all tubes used in fabrication of four-hole probe is stainless steel. The final probe has a measuring area of 1.284 mm2 . Design and fabrication details and the sizes of hypodermic tubes used for fabrication of the probes are described in Table 1.

Table 1

| Measuring tube diameter (mm) | Holding tube diameter (mm) | Probe size | | Measuring area (mm²) |
		W (mm)	T (mm)	
0.414	0.719	1.133	1.133	1.284

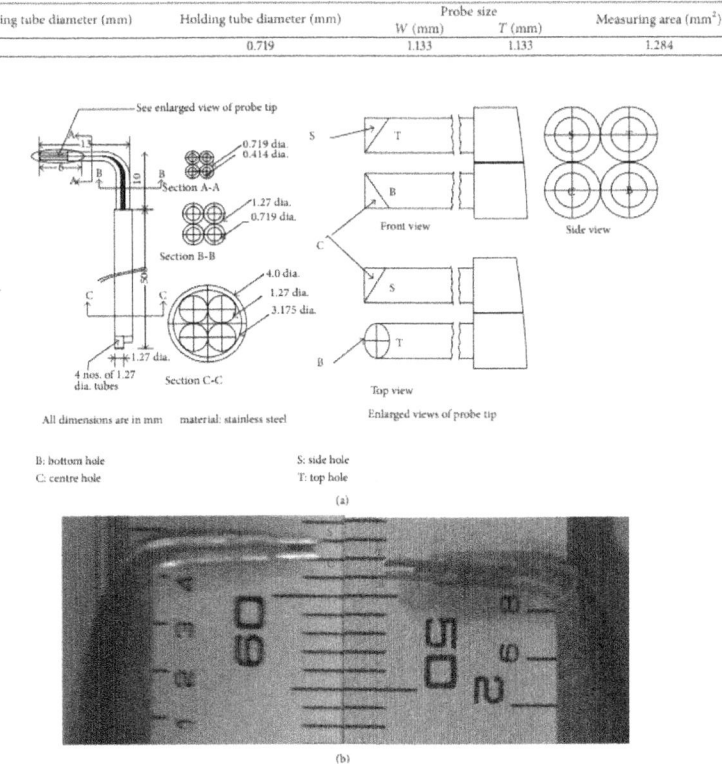

Figure 2: (a) AutoCAD drawing of probe. (b) Photo of probe tip.

The centre hole (C) with a chamfer angle of 90∘ to the tube gives a measure of total pressure and is denoted by PC which is at lower left position. The side hole is chamfered at an angle of about 35∘ to the yaw plane and is designated as P_S and is mainly sensitive to the yaw angle variation. The bottom hole (B) and top hole (T) are chamfered at an angle of about 35∘ to the pitch plane and are noted as P_B and P_T, respectively, and are mainly sensitive to the pitch angle variation. Hence four independent probe pressures are sufficient to define the threedimensional flow. It has been already demonstrated that small chamfer gives lower sensitivity and higher operating range [22]. Tip section is made of four 0.414 mm diameter tubes which are inserted into four other 0.719 mm diameter tubes. This leaves a small gap of 0.3 mm between the four 0.414 mm diameter tubes. No attempt is made to fill this gap, as it would be very difficult to do so without distorting the extremely small diameter tubes. These small gaps also seem to have very little effect on the calibration curves. All joints of small diameter tubes are made using Araldite. The 0.719 mm diameter tubes are bent at an angle of 90° with a radius of curvature of about 5 mm.

Stem Section. The stem section consists of four 1.27 mm diameter tubes of about 500 mm length which are inserted into the other ends of the 0.719 mm diameter tubes. To keep the four tubes in the proper plane they are brazed at different positions along its length. To maintain the position of the inner tubes fixed with respect to the stem, the outer tube of 3.175 mm inner diameter is also brazed with these tubes. This tube of 4 mm outer diameter acts as the probe holder.

CALIBRATION TUNNEL, DEVICE, PROCEDURE, AND PROGRAM

An open jet low speed calibration tunnel facility of Thermal Turbomachines Laboratory, Department of Mechanical Engineering, IIT Madras, is used for calibration of the miniature four-hole probe as shown in Figure 3.

Calibration device is made of base plate, c clamp, protractors, and pointers for measurement of pitch (β) and yaw (α) angles. The twenty-channel selection box and the FC012 digital micromanometer with a range of 1–200 mm of water and sensitivity of 0.1 mm of differential air pressure are used to measure the probe pressures. The micromanometer uses the output signals from the selection box to get the velocity and pressure readings.

Calibration of the probe is carried out at a velocity of 50 m/s. The probe is mounted on the probe holder with the help of sleeve such that the pressure sensing holes of the probe are to face the flow. The assembly of the probe and probe holder is kept 100 mm away from the exit of the nozzle.

Figure 3: Calibration tunnel, calibration device, probe, and instrumentation.

At first the zeroing of probe is done by setting up the pitch angle (β) to zero degree. Initially, by changing the yaw angle (α), set the position of the probe such that the pressure sensed by centre hole is maximum and the yaw angle corresponding to the maximum pressure sensed by centre hole is noted down and then the probe is rotated on both positive and negative sides of the yaw angle until the pressure at the centre hole is equal to the half of the maximum pressure sensed and the corresponding yaw angles are noted down. The mean of these yaw angles is taken as zero reference yaw angle.

After fixing the zero reference position the probe is calibrated by changing α and β in the range of −40∘ to 40∘ with an interval of 5∘ . The calibration is done by keeping α constant and by varying β. For every combination of α and β, the probe pressures are recorded.

RESULTS AND DISCUSSION

Sample Data

To determine the qualitative accuracy of the measurements, the measured pressure data is nondimensionalised with dynamic pressure and variation of various nondimensional probe pressures presented against yaw and pitch angles in Figure 4. Pressure measured by the central hole, P_C, is maximum at smaller yaw and pitch angles and varies more or less symmetrically about both yaw and pitch angles. Pressure measured by the side hole, P_S, seems to be mainly sensitive to yaw angle. Pressures measured by the bottom hole, P_B, and top hole, P_T, are mainly sensitive to pitch angle. The graphs show the expected trends of measured pressures without any abnormal values. Hence the data is found satisfactory and is used to determine calibration coefficients.

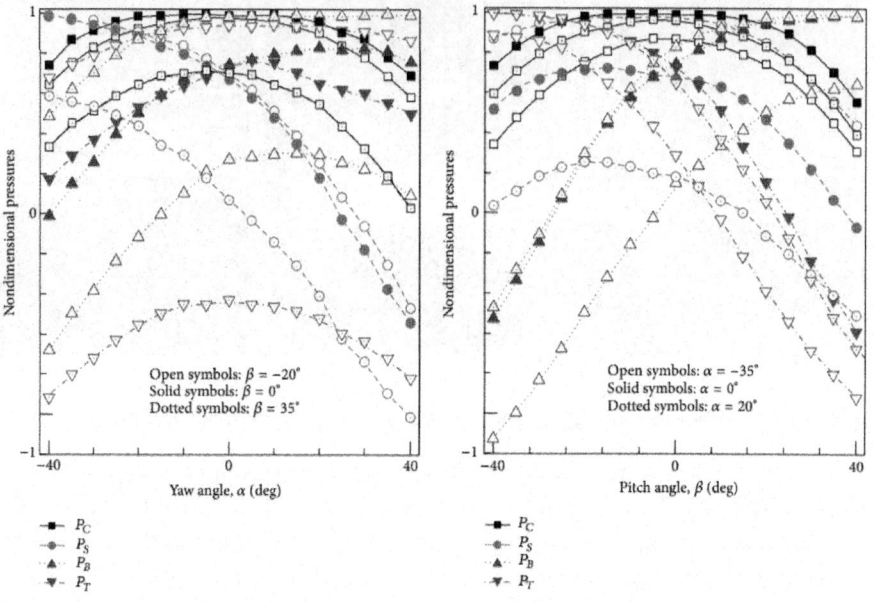

Figure 4: Sample data.

Calibration Coefficients

The calibration coefficients are defined as follows:

$$P_{BAR} = \frac{(P_S + P_B + P_T)}{3},$$

$$D = P_C - P_{BAR},$$

$$C_{PYAW} = \frac{(P_C - P_S)}{D},$$

$$C_{PPITCH} = \frac{(P_T - P_B)}{D},$$

$$C_{PTOTAL} = \frac{(P_C - P_O)}{Q},$$

$$C_{PSTATIC} = \frac{(P_{BAR} - P_{ST})}{Q},$$

$$Q = P_O - P_{ST} = P_O \quad (\text{as } P_{ST} = 0)$$

$$= \frac{(P_C - P_O)}{P_O}$$

$$= \frac{P_{BAR}}{P_O}. \tag{1}$$

Calibration Curves. The following calibration curves are presented in Figure 4. For the sake of clarity, calibration curves are shown at $10°$ intervals only.

(1) C_{PYAW} versus C_{PPITCH} for different pitch and yaw angles.

(2) C_{PTOTAL} and $C_{PSTATIC}$ contours with C_{PPITCH} and C_{PYAW} along the axis.

In Figure 4, calibration coefficients at $\alpha = -40°$ and $\beta = -40°$ are not presented. The value of probe dynamic pressure, D is very small and the magnitude of the calibration coefficients is very large.

For an ideal five-hole probe, C_{PYAW} versus C_{PPITCH} calibration curves at constant yaw angles will be horizontal and symmetrical about zero yaw angle. C_{PYAW} versus C_{PPITCH} calibration curves at constant pitch angles will be vertical and symmetrical about zero pitch angle. Because of nonlinearity in the behaviour of static pressure, these calibration curves will be curved. For a four-hole probe, the calibration curves will be asymmetric about both yaw and pitch angles at zero value. In C_{PYAW} versus C_{PPITCH} curves for $\beta = 20°$, $30°$, and $40°$, C_{PPITCH} is minimum at $\alpha = 0°$ and increases nonlinearly on both sides of the zero yaw angle. For $\beta = -20°$, $-10°$, $0°$, and $10°$, C_{PPITCH} is found to vary nonlinearly for different values of yaw angles. For $\beta = -30°$ and $-40°$, C_{PPITCH} is maximum at $\alpha = 0°$ and decreases nonlinearly on both sides of the yaw angles. At $\alpha = 0°$, C_{PYAW} is almost a straight line for various pitch angles and as α increases on both sides C_{PYAW} varies nonlinearly for different values of pitch angles as shown in Figure 4. From Figure 4, it is evident that the pitch coefficient, C_{PPITCH}, has larger sensitivity with pitch angle compared to C_{PYAW} sensitivity with yaw angle. This is expected as C_{PPITCH} depends on two pitch holes compared to C_{PYAW} which depends on one yaw hole only

with pitch angle compared to sensitivity with yaw angle. This is expected as depends on two pitch holes compared to which depends on one yaw hole only.

In the C_{PTOTAL} and $C_{PSTATIC}$ contours (Figure 5), the minimum C_{PTOTAL} contour is 0 at zero yaw and pitch angles. By definition of the C_{PTOTAL}, as the pressure sensed by the centre hole is always less than the total pressure, C_{PTOTAL} can be either zero or negative. At pitch and yaw angles away from zero values, C_{PTOTAL} becomes more negative..

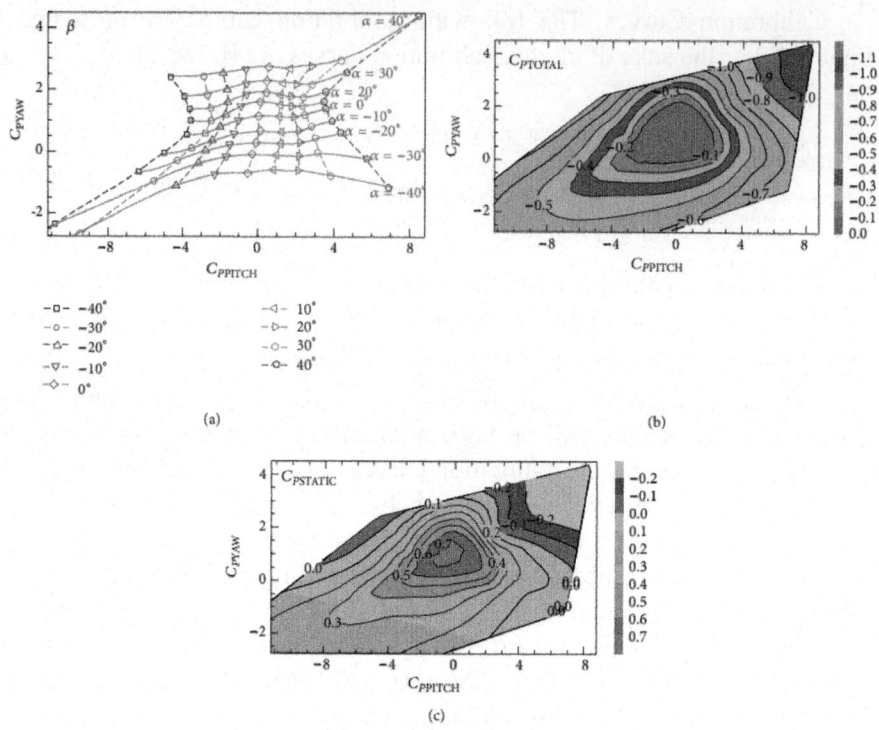

Figure 5: Calibration curves of miniature four-hole probe.

The minimum $C_{PSTATIC}$ contour is 0.7 at small yaw and pitch angles because it depends on P_{BAR} and P_{BAR} is average pressure sensed by side, bottom, and top holes. Hence the value of $C_{PSTATIC}$ decreases at higher yaw and pitch angles because P_{BAR} decreases rapidly at higher yaw and pitch angles.

Sensitivity Analysis of Calibration Coefficients

To define the accuracy of the measurements, sensitivity analysis of calibration data is carried out. The sensitivity coefficients are defined as a function of yaw or pitch angle while keeping pitch or yaw angle constant. The sensitivity coefficients are shown in Figure 6.

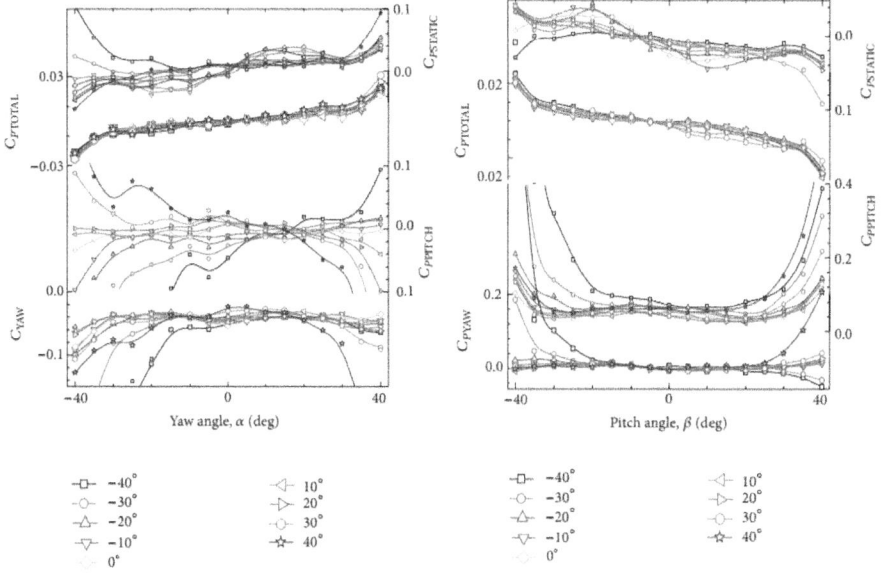

Figure 6: Sensitivity curves of calibration coefficients.

Sensitivity coefficients are defined as

$$\Delta C_{P_i} = \frac{\left(C_{P_{(i+1)}} - C_{P_{(i-1)}}\right)}{\left(\text{Angle}\,(i+1) - \text{Angle}\,(i-1)\right)},$$

(2)

where C_p is the one of the four calibration coefficients, namely C_{PYAW}, C_{PPITCH}, C_{PTOTAL}, and C_{PSTATIC}, and i is the yaw or pitch angle where the calibration data is taken.

The probe pressures change rapidly at large yaw and pitch angles. Therefore calibration coefficients at large values of yaw and pitch angles have higher sensitivity. Although higher sensitivity implies more accurate measurements but operating range of the probe will be less. It is to be kept in mind that small error in pressures results in large errors in calibration coefficients and their sensitivity. At low values of yaw and pitch angles, sensitivity coefficients are low. The probe hole chamfered angles are small, about 35°. It is already demonstrated that small chamfer angles result in lower sensitivity and large chamfer angles result in higher sensitivity [22].

Validation of Calibration Data

Sitaram and Kumar [23] developed a look up table method to determine the four unknown quantities, namely, yaw and pitch angles and static and total pressure coefficients from the calculated yaw and pitch coefficients of a five-hole probe. The yaw and pitch coefficients are calculated from the measured pressure data. The same method is utilised for determining the flow quantities from the present four-hole probe measurements. No additional data is taken for interpolation during the calibration of the probe. However the calibration data at intervals of 10° rather than 5° are used. All the calibration data are used as measured data. A calibration interval of 10° is large. Sumner [24] recommended that this is the largest calibration interval that can be used with a seven-hole probe.

The interpolated values are compared with those obtained during calibration. Histograms of errors in yaw and pitch angles are presented in Figure 7. The errors at the extremes of the calibration range, that is, ±40° of yaw and pitch angles, are omitted from these graphs. Most of the errors in yaw and pitch angles are within ±1° and most of the errors in total, static, and dynamic pressures are within −0.005 to 0.01% of the dynamic head.

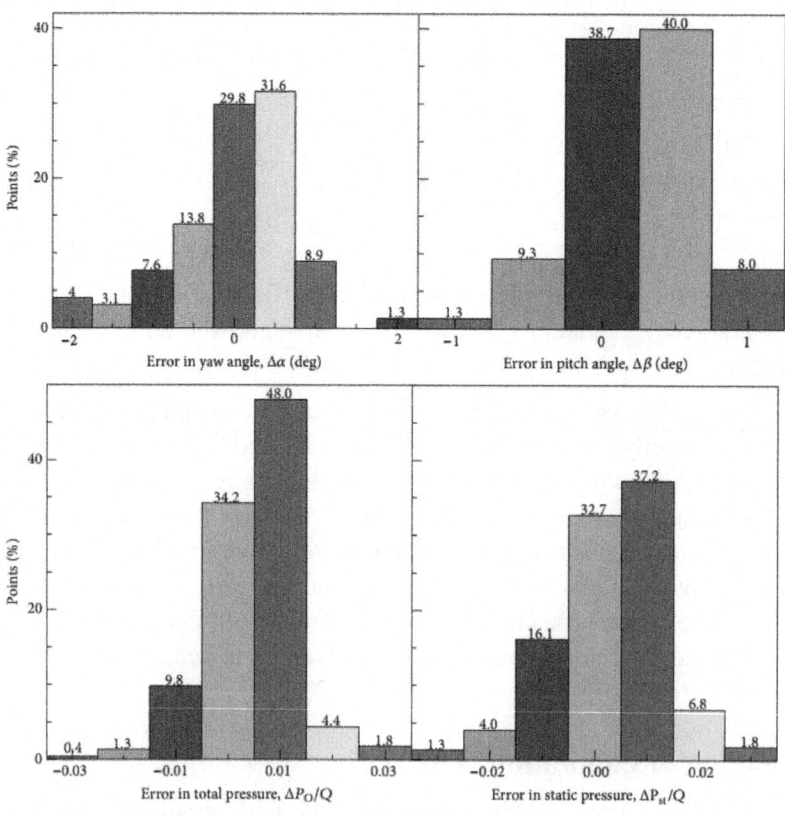

Figure 7: Histograms of interpolated errors in percentages.

The maximum absolute, average, and rms values of errors in yaw and pitch angles and total, static, and dynamic pressures are also presented in Table 2.

Table 2

Parameter	Maximum error	Minimum error	Average value	RMS value
$\Delta\alpha$	1.7°	−2.4°	−0.26°	0.80°
$\Delta\beta$	1.3°	−1.1°	0.01°	0.42°
$\Delta P_O/Q$	0.034	−0.031	−0.001	0.008
$\Delta P_{ST}/Q$	0.026	−0.034	−0.002	0.012
$\Delta Q/Q$	0.029	−0.049	0.003	0.011

Except for yaw angle and dynamic pressure, the errors are very small. The large values of error occur near the extreme range of calibration. The errors are due to the data reduction program only. All other measurement errors such as instrumentation errors, errors due calibration (zero angle settings, pitch and yaw angle measurements during calibration, etc.), are not included. The calibration data is given in 10 deg. interval. The calibration data in 5 deg. interval (excluding the data at 10 deg. interval) are given as measured data. The errors are almost negligible when both calibration data and measured data are given in 5 deg. interval. The errors presented by Lee and Jun [7], who used a calibration interval of 5° in their data reduction program, have similar magnitude.

Comparison of Measurements in a Calibration Duct

The four-hole probe along with a conventional truncated conical head with perpendicular holes (3 mm dia. head) and an extremely small Pitot tube is used to measure the flow across the calibration section of the calibration tunnel available at Thermal Turbomachines Laboratory, Department of Mechanical engineering, IIT Madras (0.5 mm tip dia. of 5 mm length; this tube is extended to 0.8 mm dia. tube, followed by 1.27 mm dia. tube to reduce the response time). A schematic of the calibration tunnel is shown in Figure 8.

The circumferentially averaged wall static pressure of the settling chamber is equal to the total pressure in the calibration section. The static pressure at the measurement station is measured by means of four circumferentially averaged wall static pressure taps. The probes are traversed from the centre of the calibration section to the end of the opposite wall. A manual traversing mechanism with 1 mm measurement resolution along the radial direction and 1° measurement resolution in the yaw plane is used to traverse the probes. The three probes are nulled at the centre of the calibration section and traversed in large intervals (10 mm) near the centre. As the probes approach the opposite wall, the intervals are reduced to 5 mm, 2 mm, and 1 mm. The results of these measurements are presented in Figure 9.

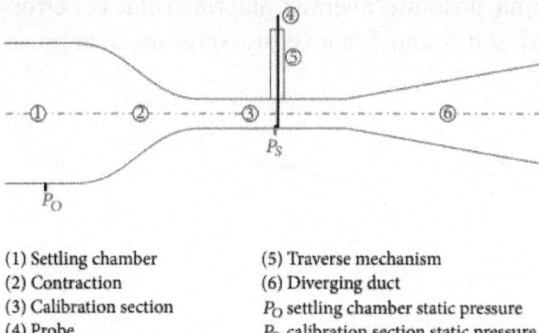

(1) Settling chamber   (5) Traverse mechanism
(2) Contraction   (6) Diverging duct
(3) Calibration section P_O settling chamber static pressure
(4) Probe P_S calibration section static pressure

Figure 8: Schematic of calibration tunnel.

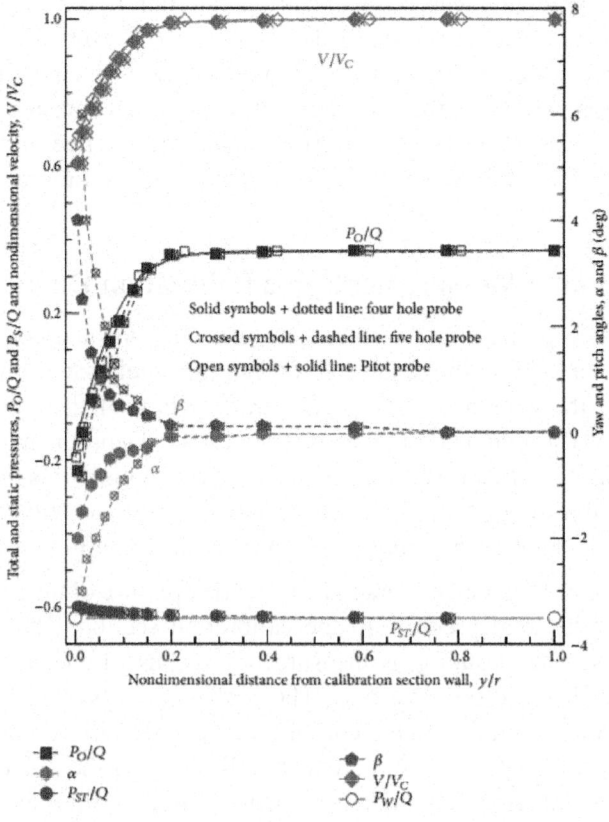

Figure 9: Comparison of probe measurements in the calibration section of the calibration tunnel.

From Figure 9, it is evident that the nondimensional total pressure and velocity measured by all the three probes are in good agreement at the centre and away from the centre up to a distance of 0.2 times radius from the calibration section wall. For the Pitot tube, wall static pressure is used to calculate velocity. The flow is uniform in this region. The thickness of the boundary layer is about 0.2 times the radius of the calibration section. The nondimensional static pressures measured by the four-hole and five-hole probes are in good agreement with the nondimensional static pressure measured by the wall static pressure taps. Only very near the wall of the calibration section, the static pressures measured by the probes are slightly higher than that measured by the wall static pressure taps.

Near the wall, pressure probes suffer from two major sources of errors, namely, pressure or velocity gradients and surface proximity errors. In addition, the turbulence intensity increases as the probes approach the walls. Recent investigations [25] have shown that the turbulence intensity affects the calibration characteristics of multihole probes and suggested that the multihole probes should be calibrated at the same turbulence intensity of the flow to be measured. Sitaram et al. [26] have discussed these and other sources of errors in the application of five-hole probes and they gave estimates of their magnitudes. Near the wall, the flow is modified due to the presence of the probe and the probe gives erroneous results. These erroneous results are usually confined to a distance of about twice the diameter of the probe. All the parameters that are measured, namely, total and static pressures, flow angles, and velocity and its three components, are affected due to the surface proximity. However, usually static pressure and pitch angle show larger errors. As the tip diameter of the Pitot tube is only 0.5 mm, it can measure total pressure accurately 1 mm away from the wall. The nominal sizes of the four-hole and five-hole probes are 1.1 mm and 3 mm, respectively, and the errors measured by these probes extend up to distances of about 2 and 6 mm, respectively. This can be clearly demonstrated by examination of the radial distribution of yaw and pitch angles.

The radial distribution of yaw and pitch angles measured by the four-hole and five-hole probes is also presented in Figure 9. At the centre of the calibration section, both yaw and pitch angles are zero as the flow is one-dimensional and aligned with the axis of the calibration section. These angles are close to zero up to very close to the wall. The small discrepancies can be attributed to the interpolation errors. As expected, the angles measured by the four-hole probe are nonzero very close to the wall (). For the five-hole probe, the angles measured are nonzero and are up to /. The maximum values of yaw angles are −2 and −3, for the four-hole and five-hole probes, respectively. The maximum values of pitch angles are 4 and 6, respectively, for the four-hole and five-hole probes.

CONCLUSIONS

A miniature four-hole probe with minimum spatial error is designed and fabricated. The probe can be used to analyze three-dimensional flows with large pressure and velocity gradients in all directions as in cases of tip clearance vortex, flows in corner of blades, end wall flows, and other complex flows that occur in turbomachinery. The probe is calibrated in the range of −40° to 40° in both yaw and pitch planes with an interval of 5°. Calibration coefficients are defined, determined, and plotted. Sensitivity analysis of the calibration data is also performed. A lookup table method is used to interpolate the four unknown quantities, namely, total and static pressures and flow angles. The maximum absolute errors in yaw and pitch angles are 2.4 and 1.3 deg. respectively. The maximum absolute errors in total, static and dynamic pressures are 3.4, 3.9 and 4.9% of the dynamic pressures respectively. Measurements made with this probe, a conventional five-hole probe and a miniature Pitot probe across a calibration section, demonstrated that the errors due to gradient and surface proximity for this probe are considerably reduced compared to the five-hole probe. Hence, this probe is more suitable to measure three-dimensional flows with large pressure and velocity gradients as in cases of tip clearance vortex, flows in corner of blades, end wall flows, and other complex flows that occur in turbomachinery.

ACKNOWLEDGMENTS

The authors would like to thank Mr. M. Veeraraghavan and Mr. N. Giri of Thermal Turbomachines Laboratory, Department of Mechanical Engineering, IIT Madras, for fabrication of the miniature four-hole probe. The authors would like to thank Mr. R. Bharath Viswanath, Project Associate, Thermal Turbomachines Laboratory, Department of Mechanical Engineering, IIT Madras, for his help in running lookup table method program. The authors would like to thank the reviewers for their suggestions, which improved the quality of the paper substantially.

REFERENCES

1.   L. Treaster and A. M. Yocum, "The calibration and application of five-hole probes," nstrumentation Society of America Transactions, vol. 18, no. 3, pp. 23–34, 1979.

2.   J. Pisasale and N. A. Ahmed, "A novel method for extending the calibration

range of five-hole probe for highly three-dimensional flows," Flow Measurement and Instrumentation, vol. 13, no. 1-2, pp. 23–30, 2002.

3. Pissasale and N. A. Ahmed, "Theoretical calibration for highly three-dimensional low-speed flows of a five-hole probe," Measurement Science and Technology, vol. 13, no. 7, pp. 1100–1107, 2002.

4. Pissasale and N. A. Ahmed, "Development of a functional relationship between port pressures and flow properties for the calibration and application of multihole probes to highly three-dimensional flows," Experiments in Fluids, vol. 36, no. 3, pp. 422–436, 2004.

5. T. Yasa and G. Paniagua, "Robust procedure for multi-hole probe data processing," Flow Measurement and Instrumentation, vol. 26, pp. 46–54, 2012.

6. R. G. Dominy and H. P. Hodson, "Investigation of factors influencing the calibration of five-hole probes for three-dimensional flow measurements," Journal of Turbomachinery, vol. 115, no. 3, pp. 513–519, 1993.

7. S. W. Lee and S. B. Jun, "Reynolds number effects on the non-nulling calibration of a cone-type five-hole probe for turbomachinery applications," Journal of Mechanical Science and Technology, vol. 19, no. 8, pp. 1632–1648, 2005.

8. D. Telionis, Y. Yang, and O. Rediniotis, "Recent developments in multi-hole probe technology," inProceedings of 20th International Congress of Mechanical Engineering (COBEM ‹09), Gramado, Brazil, November 2009.

9. S. J. Lien and N. A. Ahmed, "An examination of suitability of multi-hole pressure probe technique for skin friction measurement in turbulent flow," Flow Measurement and Instrumentation, vol. 22, no. 3, pp. 153–164, 2011.

10. P. M. Ligrani, B. A. Singer, and L. R. Baun, "Miniature five-hole pressure probe for measurement of three mean velocity components in low-speed flows," Journal of Physics E: Scientific Instruments, vol. 22, no. 10, pp. 868–876, 1989.

11. V. Chernoray and J. Hjärne, "Improving the accuracy of multihole probe measurements in velocity gradients," in Proceedings of the ASME Turbo Expo, pp. 125–134, Berlin, Germany, June 2008.

12. J. Town, A. Akturk, and C. Camci, "Total pressure correction of a sub-miniature five-hole probe in areas of pressure gradients," in Proceedings of the ASME Turbo Expo 2012: Turbine Technical Conference and Exposition (GT ‹12), pp. 855–861, Copenhagen, Denmark, June 2012.

13. H. T. Hoenen, R. Kunte, P. Waniczek, and P. Jeschke, "Measuring failures and correction methods for pneumatic multihole probes," in Proceedings of the ASME Turbo Expo : Turbine Technical Conference and Exposition (GT ‹12), pp. 721–729, Copenhagen, Denmark, June 2012.

14. J. R. Erwin, "Experimental techniques: instrumentation," in High Speed Aerodynamics and Jet Propulsion, vol. 5 of Aerodynamics of Turbines and Compressors, p. 173, 1964.

15. N. K. Maheshwari, D. Saha, R. K. Sinha, and V. Venkat Raj, Velocity Measurement in the Three Dimensional Flow Fluid in a Nuclear Reactor Calandria Using Four Hole Probe, Reactor Engineering Division, Bhabha Atomic Research Centre, Mumbai, India, 1998.

16. AC-flow., http://ac-flow.com/upload/files/4-hole%20wedge%20 type%20probe.pdf.

17. Heneka, "Instantaneous three dimensional flow measurements with a four hole wedge probe," inProceedings of the 7th Symposium on Measuring Techniques for Transonic and Supersonic Flow in Cascades and Turbomachines, Aachen, Germany, 1983.

18. R. W. Ainsworth, J. L. Allen, and J. J. M. Batt, "The development of fast response aerodynamic probes for flow measurements in turbomachinery," ASME Journal of Turbomachinery, vol. 117, no. 4, pp. 625–634, 1995.

19. J. P. Schlienger, Evolution of unsteady secondary flows in a multistage shrouded axial turbine [dissertation], ETH, Zürich, Switzerland, 2003, ETH no. 15230.

20. C. Shepherd, "A four hole pressure probe for fluid flow measurements in three dimensions," Journal of Fluids Engineering, vol. 103, no. 4, pp. 590–594, 1981.

21. N. Sitaram and A. L. Treaster, "A simplified method of using four-hole probes to measure three dimensional flow fields," Journal of Fluids Engineering, Transactions of the ASME, vol. 107, no. 1, pp. 31–35, 1985.

22. N. Sitaram and K. Srikanth, "Effect of chamfer angle on the calibration curves of five hole probes," inProceedings of the 39th National Conference on Fluid Mechanics and Fluid Power, Paper No. FMFP2012-240, p. 10, SVNIT, Surat, India, December 2012.

23. N. Sitaram and S. Kumar, "Look up table method for five hole probe data reduction," in Proceedings of the 38th National Conference on Fluid Mechanics and Fluid Power, vol. 8 of Paper No. EM05, MANIT, Bhopal, India, December 2011.

24. D. Sumner, "A comparison of data-reduction methods for a seven-hole probe," Journal of Fluids Engineering, vol. 124, no. 2, pp. 523–527, 2002.

25. Crowley, I. I. Shinder, and M. R. Moldover, "The effect of turbulence on a multi-hole Pitot calibration,"Flow Measurement and Instrumentation, vol. 33, pp. 106–109, 2013.

26. N. Sitaram, B. Lakshminarayana, and A. Ravindranath, "Conventional probes for the relative flow measurement in a turbomachinery rotor blade passage," ASME Journal of Engineering for Gas Turbines and Power, vol. 103, no. 2, pp. 406–414, 1981.

CITATION

CHAPTER 1

Ariavie, Go; Oyekale Jo Emagbetere E, Performance Modelling of Steam Turbine Performance using Fuzzy Logic Membership Functions, ISSN 1119-8362.

CHAPTER 2

Chih-Neng Hsu, "A Study on Fluid Self-Excited Flutter and Forced Response of Turbomachinery Rotor Blade," Mathematical Problems in Engineering, vol. 2014, Article ID 437158, 20 pages, 2014. doi:10.1155/2014/437158.

CHAPTER 3

R A Van den Braembussche, Challenges and progress in turbomachinery design systems, doi:10.1088/1757-899X/52/1/012001.

CHAPTER 4

X X Huang, E Egusquiza, C Valero and A Presas, Dynamic behaviour of pump-turbine runner: From disk to prototype runner, doi:10.1088/1757-899X/52/2/022036.

CHAPTER 5

Cheng Xu and Ryoichi S. Amano, "Empirical Design Considerations for Industrial Centrifugal Compressors," International Journal of Rotating Machinery, vol. 2012, Article ID 184061, 15 pages, 2012. doi:10.1155/2012/184061.

CHAPTER 6

Roberto Capata and Enrico Sciubba, Experimental Fitting of the Re-Scaled Balje Maps for Low-Reynolds Radial Turbomachinery, doi:10.3390/en8087986.

CHAPTER 7

M. Suzuki, "Numerical Analysis of Horizontal-Axis Wind Turbine Characteristics in Yawed Conditions," Open Journal of Fluid Dynamics, Vol. 2 No. 4A, 2012, pp. 331-336. doi: 10.4236/ojfd.2012.24A041.

CHAPTER 8

F. R. Menter and R. B. Langtry (2012). Transition Modelling for Turbomachinery Flows, Low Reynolds Number Aerodynamics and Transition, Dr. Mustafa Serdar Genc (Ed.), ISBN: 978-953-51-0492-6, InTech, DOI: 10.5772/38675.

CHAPTER 9

Jason Town and Cengiz Camci, "A Time Efficient Adaptive Gridding Approach and Improved Calibrations in Five-Hole Probe Measurements," International Journal of Rotating Machinery, vol. 2015, Article ID 376967, 14 pages, 2015. doi:10.1155/2015/376967.

CHAPTER 10

Reid A. Berdanier and Nicole L. Key, "Experimental Investigation of Factors Influencing Operating Rotor Tip Clearance in Multistage Compressors," International Journal of Rotating Machinery, vol. 2015, Article ID 146272, 13 pages, 2015. doi:10.1155/2015/146272.

CHAPTER 11

Fadhl M. Al-Akwaa (2012). Analysis of Gene Expression Data Using Biclustering Algorithms, Functional Genomics, Dr. Germana Meroni (Ed.), ISBN: 978-953-51-0727-9, InTech, DOI: 10.5772/48150.

CHAPTER 11

Ravirai Jangir, Nekkanti Sitaram, and Ct Gajanan, "A Miniature Four-Hole Probe for Measurement of Three-Dimensional Flow with Large Gradients," International Journal of Rotating Machinery, vol. 2014, Article ID 297861, 12 pages, 2014. doi:10.1155/2014/297861.

INDEX